Intelligent Systems Reference Library

Volume 53

Series Editors

Janusz Kacprzyk, Polish Academy of Sciences, Warsaw, Poland
e-mail: kacprzyk@ibspan.waw.pl

Lakhmi C. Jain, University of Canberra, Canberra, Australia
e-mail: Lakhmi.Jain@unisa.edu.au

For further volumes:
http://www.springer.com/series/8578

About this Series

The aim of this series is to publish a Reference Library, including novel advances and developments in all aspects of Intelligent Systems in an easily accessible and well structured form. The series includes reference works, handbooks, compendia, textbooks, well-structured monographs, dictionaries, and encyclopedias. It contains well integrated knowledge and current information in the field of Intelligent Systems. The series covers the theory, applications, and design methods of Intelligent Systems. Virtually all disciplines such as engineering, computer science, avionics, business, e-commerce, environment, healthcare, physics and life science are included.

Vijay Kumar Mago · Vahid Dabbaghian
Editors

Computational Models of Complex Systems

Springer

Editors
Vijay Kumar Mago
Department of Computer Science
Troy University
Troy
USA

Vahid Dabbaghian
The MoCSSy Program,
The IRMACS Centre
Simon Fraser University
Burnaby
Canada

Vijay Kumar Mago started this work when he was a postdoctoral fellow at the IRMACS Centre, Simon Fraser University, Canada

ISSN 1868-4394 ISSN 1868-4408 (electronic)
ISBN 978-3-319-34825-4 ISBN 978-3-319-01285-8 (eBook)
DOI 10.1007/978-3-319-01285-8
Springer Cham Heidelberg New York Dordrecht London

© Springer International Publishing Switzerland 2014
Softcover reprint of the hardcover 1st edition 2014
This work is subject to copyright. All rights are reserved by the Publisher, whether the whole or part of the material is concerned, specifically the rights of translation, reprinting, reuse of illustrations, recitation, broadcasting, reproduction on microfilms or in any other physical way, and transmission or information storage and retrieval, electronic adaptation, computer software, or by similar or dissimilar methodology now known or hereafter developed. Exempted from this legal reservation are brief excerpts in connection with reviews or scholarly analysis or material supplied specifically for the purpose of being entered and executed on a computer system, for exclusive use by the purchaser of the work. Duplication of this publication or parts thereof is permitted only under the provisions of the Copyright Law of the Publisher's location, in its current version, and permission for use must always be obtained from Springer. Permissions for use may be obtained through RightsLink at the Copyright Clearance Center. Violations are liable to prosecution under the respective Copyright Law.
The use of general descriptive names, registered names, trademarks, service marks, etc. in this publication does not imply, even in the absence of a specific statement, that such names are exempt from the relevant protective laws and regulations and therefore free for general use.
While the advice and information in this book are believed to be true and accurate at the date of publication, neither the authors nor the editors nor the publisher can accept any legal responsibility for any errors or omissions that may be made. The publisher makes no warranty, express or implied, with respect to the material contained herein.

Printed on acid-free paper

Springer is part of Springer Science+Business Media (www.springer.com)

Dedicated to Dr. Peter Borwein for his contribution to research and education on his sixtieth birthday

Acknowledgments

We are extremely grateful to the IRMACS Center at Simon Fraser University, Burnaby, Canada for the administrative, and technical support that greatly facilitated this research.

We also wish to thank the reviewers for their constructive comments on the chapters.

Laurens Bakker	KPMG, Netherlands
Nitin Bhatia	DAV College, Jalandhar, India
Gabriel Burstein	City University of New York, New York, USA
Thomas Couronné	Orange Labs/France Télécom R&D, France
Suzana Dragićević	Simon Fraser University, Burnaby, Canada
Richard Frank	Simon Fraser University, Burnaby, Canada
Philippe Giabbanelli	Simon Fraser University, Burnaby, Canada
Piper Jackson	Simon Fraser University, Burnaby, Canada
Manpreet Jammu	Microsoft, Washington, USA
Mamta Khosla	Dr. B. R. Ambedkar National Institute of Technology, Jalandhar, India
Vijay K Mago	Simon Fraser University, Burnaby, Canada
Bimal Kumar Mishra	Birla Institute of Technology, Ranchi, India
Susan Mniszewski	Los Alamos National Laboratory, USA
Elpiniki I. Papageorgiou	Technological Educational Institute of Lamia, Greece
Laurent Tambayong	University of California, Irvine, USA
Pietro Terna	University of Torino, Italy
Katie Wuschke	Simon Fraser University, Burnaby, Canada

Contents

Chapter 1
Computational Models of Complex Systems: An Introduction

Vijay Kumar Mago and Vahid Dabbaghian

Abstract Focused research on understanding the complexities of social and economic systems through computational model is the main theme of this book. Many new applications of computational models like Kabbalah Tree of Life, Spectral Analysis, Fuzzy Cognitive Maps, Artificial Neural Network, Cellular Automata, and Agent based modelling approaches are proposed and applied successfully to investigate intricacies of the systems.

1 Introduction

Complex networks exist throughout the domains of nature and society: swarms, circulatory systems, roads, power grids. These networks enable the efficient distribution of resources, resulting in greater and more impressive activity. Social systems are networks that go one step further: they are not only there for the distribution of resources, but also to act as a medium for the interaction between numerous intelligent entities. Thus they combine efficient resource use with intense productivity, in the sense that interactions between these entities produce numerous effects on the system, for better or for worse [1]. We live our lives in a nexus of numerous social systems: family, friends, organizations, nations. We benefit from the fruits of their power, including energy, learning, wealth, and culture. We also struggle with the crises they generate, such as crime, war, pollution, and illness. Our motivation for studying these systems is clear: they are the fabric upon which our lives are woven.

V. K. Mago (✉) · V. Dabbaghian
The Modelling of Complex Social Systems, The IRMACS Centre,
Simon Fraser University, Burnaby, Canada
e-mail: vmago@sfu.ca

V. Dabbaghian
e-mail: vdabbagh@sfu.ca

V. K. Mago and V. Dabbaghian (eds.), *Computational Models of Complex Systems*,
Intelligent Systems Reference Library 53, DOI: 10.1007/978-3-319-01285-8_1,

© Springer International Publishing Switzerland 2014

Computational models provide us with opportunities not matched in the real world. They allow us to apply our best analytical methods to subject problems defined in a clearly mathematical manner. They give us a chance to test our designs in full before committing expensive resources to their production. They isolate a subject from an unbounded environment, allowing controllable experimentation not possible in a messy world. Computation has been posited as a third pillar of science, alongside theory and empirical research [4]. For example, simulation of computational models allows the testing of theories in a manner that is both fundamentally deductive and experimental in nature [2]. This volume provides many examples of how computational modelling has been used to in useful and innovative ways. Building up a rich foundation of literature [3], these models span a wide variety of application fields, making evident the importance of computational modelling to complex systems science.

2 New Contributions

In Chap. 2, Burstein and Negoita begin the volume by considering a model based in religious philosophy of the integrated aspects of human existence. This model, the Kabbalah Tree of Life, integrates an understanding of emotion and cognition, and is thus provides valuable insight into behavioural economics and finance. The authors show how traditional knowledge can be combined with modern computational research. Other models related to finance follow. In Chap. 3, Gray and McManus use spectral analysis to explain the robustness of the Brazilian and Australian economies, despite uncertainty in equity markets. Next, in Chap. 4, Cervelló-Royo et al. present a metric for predicting confidence in the unstable and highly connected realm of European economies. They also employ a diffusion model to study the dynamics of this metric due to both local and international interactions. As can be seen in the works, the highly numerical but complex nature of economic activity makes computational modelling an appealing choice of research method.

Healthcare is another field for which computational modelling has become recognized as an effective tool [6–8]. Fall incidents at long-term care facilities are of significant concern due to their effect on patient mortality, but simple models are incapable of including the many factors that can lead to a fall. In Chap. 5, Mago et al. employ the Fuzzy Cognitive Map technique in order to model such falls. The result is a model useful for conceptualizing, analyzing, and communicating the subject phenomena. In Chap. 6, Kalita et al. employ a different approach, Artificial Neural Networks (ANN), for a different healthcare modelling task: the classification of child disabilities, similar to the one proposed in [5]. This approach is a good fit for the problem, since ANNs are adaptable to complex data regardless of application field. These examples show how computational modelling provides assistance for dealing with serious problems, and can result in both cost-savings and reduced suffering.

Healthcare applications are not limited to the realm of software, as modelling can also be useful in the development of medical hardware. In Chap. 7, Arjunan et al. present their research into the potential of using technology to read the restricted movements of people who have suffered neuro-trauma in order to allow them to use computers. In particular, they consider how the detection and analysis of facial movements related to talking can be used to understand patients who are no longer capable of speech. In Chap. 8, Jaggernauth models the social side of bio-medical device adoption for healthcare of older adults. Computational modelling is used here to include personal considerations of the individual user when thinking about a given medical device, particularly with regards to the manufacturing process.

In Chap. 9. Aagesen and Dragicevic use a variety of modelling techniques to look at land use change in early agricultural communities. The overlapping of geographical features and social interaction allow for the combined use of geographical information systems (GIS), cellular automata (CA), and agent-based modelling (ABM). Another example of modelling of a geographical nature can be seen in Chap. 10, Tambayong and White use a variety of mathematical and statistical modelling techniques to compare the features of human settlements over a large historical period. In particular, they examine the relationship between external interactions (such as trade and conflict) and internal characteristics (such as productivity). Again, in this research we see the fascinating overlapping of geographical features with social networks that can lead to an interesting modelling approach.

Security is the final subject of computational modelling covered in this book. In Chap. 11, Frank et al. present a method for identifying terrorist websites on the internet based on keyword analysis. Of particular importance is the capacity for their system to distinguish between terrorist websites and anti-terrorist websites, despite similar vocabulary usage. In Chap. 12, Pina-Garcia and Gu look at the detection of emerging threats in social networks. Connections between members of a social network constitute a social geography, which can be traversed algorithmically in order to reveal information relevant to a given problem under consideration. The authors apply a modelling technique (random walks) typically applied to biological simulation, and explain the relevance of this method to the realm of human sociality.

3 Conclusion

This volume presents a selection of works in a variety of fields showing the value of computational modelling as a research tool when investigating complex systems. We see modelling being used as tool to answer both theoretical and practical questions. This new wave of science has been enabled by advances in computing technology, but its ongoing impact on a variety of sciences and related disciplines is manifold and pervasive. Computational modelling is capable of capturing, manipulating, and communicating knowledge in ways not previously possible. Successes in both experimental results and methodology feed into each other and further reinforce a foundation of knowledge and practice upon which to base evolving vectors of research.

The selection of works presented here each seek to meaningfully advance science in their own way, and it is our sincere hope that you will agree that they are valuable and interesting contributions to this dynamic and growing area of inquiry.

References

1. Bettencourt, L.M., Lobo, J., Helbing, D., Kühnert, C., West, G.B.: Growth, innovation, scaling, and the pace of life in cities. Proc. Natl. Acad. Sci. **104**(17), 7301–7306 (2007)
2. Epstein, J.M.: Agent-based computational models and generative social science. In: Generative Social Science: Studies in Agent-Based Computational Modeling, pp. 4–46. Princeton University Press, Princeton (1999)
3. Gilbert, G.N.: Computational Social Science. Sage, London (2010)
4. Hummon, N.P., Fararo, T.J.: The emergence of computational sociology. J. Math. Sociol. **20** (2–3), 79–87 (1995)
5. Mago, V.K., Syamala Devi, M., Bhatia., A., Mehta, R.: Multi-agent systems for healthcare simulation and modeling: applications for system improvement, chap. In: Using Probabilistic Neural Network to Select a Medical Specialist Agent, pp. 164–177. Medical Information Science Reference, Hershey (2010)
6. Mago, V.K., Mago, A., Sharma, P., Mago, J.: Fuzzy logic based expert system for the treatment of mobile tooth. In: Software Tools and Algorithms for Biological Systems, pp. 607–614. Springer, Heidelberg (2011)
7. Mago, V.K., Bhatia, N., Bhatia, A., Mago, A.: Clinical decision support system for dental treatment. J. Comput. Sci. **3**(5), 254–261 (2012)
8. Mago, V., Mehta, R., Woolrych, R., Papageorgiou, E.: Supporting meningitis diagnosis amongst infants and children through the use of fuzzy cognitive mapping. BMC Med. Inform. Decis. Making **12**(1), 98 (2012)

Chapter 2
A Kabbalah System Theory Modeling Framework for Knowledge Based Behavioral Economics and Finance

Gabriel Burstein and Constantin Virgil Negoita

Abstract Kabbalah and its Tree of Life integrate the cognitive, behavioral/emotional and action levels of human existence, explaining the relations between these and their unity. This makes it an ideal framework for behavioral economics and finance where cognitive and emotional biases and heuristics play a central role. In Burstein and Negoita [9], we began to develop a Kabbalah system theory, modeling the Tree of Life as a hierarchical three level feedback control system corresponding to the cognitive, behavioral/emotional and action levels. We will further develop this here by focusing on system dynamics, in order to create a Kabbalah system theory modeling framework for a knowledge based behavioral economics and finance. In this new framework, emotional intelligence theory [19–21, 40] and Polanyi personal tacit/explicit knowledge theory [36, 37] are used to model the emotional and cognitive level processes. Here we are connecting these theories with behavioral economics and finance. While behavioral finance focuses on the impact of knowledge/cognition and emotional factors, the intrinsic dynamics of these factors is not considered in depth. This is why behavioral economics and finance can benefit from integrating emotional intelligence and knowledge theory in their modeling. Our Kabbalah system theory creates the modeling framework for that. In particular, it allows connecting the recent knowledge based economic theory attempts with behavioral economics in a unified knowledge based behavioral economics. We apply this in particular to Kabbalah behavioral, knowledge based asset pricing modeling.

G. Burstein (✉) · C. V. Negoita
Department of Computer Science, Hunter College, City University of New York, New York, USA
e-mail: gabrielburstein@yahoo.com

V. K. Mago and V. Dabbaghian (eds.), *Computational Models of Complex Systems*, Intelligent Systems Reference Library 53, DOI: 10.1007/978-3-319-01285-8_2,
© Springer International Publishing Switzerland 2014

1 Introduction and Outline

One of the primary causes of the present economic and financial crises we are witnessing now is the failure of the applied classical economics and finance to address in practice in a unified way: (1) the personal knowledge and cognitive level, (2) the emotional and behavioral level and (3) the action level of economic and financial dynamics.

There is now considerable progress of modern economic and financial theory in addressing how behavioral biases and heuristics due to levels 1 and 2 affect processes at level 3 [9, 21, 22, 36, 39, 40]. However, these do not include models for the intrinsic mechanisms at knowledge and emotional/behavioral levels but rather focus on the impact of those levels on economics and finance.

The knowledge theory of Polanyi [34, 35] and emotional intelligence theory of Goleman [17–19] and Salovey and Mayer [38] provide ideal frameworks for the dynamics at levels 1 and 2.

An interdisciplinary holistic, system theoretic framework is needed to integrate these together. At the present, modeling approaches in behavioral economics and finance are predominantly statistical or empiric, reductionist rather than holistic integrative and system theoretic. In order to approach simultaneously levels 1, 2, 3, one needs a holistic integrative common modeling language that is both quantitative and structural qualitative. This is exactly our objective here: a holistic Kabbalah system theory modeling framework for knowledge and behavioral based economics and finance, integrating knowledge theory and emotional intelligence.

General system theory (GST) and cybernetics [42–44] emerged as a program to address in a holistic way all aspects of different type of systems and the interdependence between these in a unified formalism. System theory has been so far very little applied to economics and finance. Lange was among the first one attempting to formulate an "economic cybernetics" [23].

A new type of system theory models are required to approach the different quantitative and structural qualitative natures of multi-faceted complex systems involving the human element. In the first part of this chapter we are introducing our Kabbalah based system theory that we began developing in [6] in order to address simultaneously in a unified framework the triple nature of economic and financial systems. We will then focus here on how to apply it to behavioral economics and finance and this will require developing new system dynamics modeling introduced here for the first time.

Kabbalah and its Tree of Life integrate together the cognitive, emotional/behavioral and action levels of human existence and is thus an ideal integrative framework for knowledge based behavioral economics and finance. The Tree of Life actually has three interconnected levels and they are exactly the cognitive, the emotional/behavioral and the action level.

We are first presenting basic elements of Kabbalah, the ancient philosophical and scientific analysis of creation and existence, a truly scientific thinking developed across many centuries by Rabbis Shimon Bar Yochai (Rashbi), Isac Luria

(Ari or Arizal), Moshe Cordovero (Ramak), Chayyim Vital, Moshe Hayim Luzzatto (Ramhal), Yehuda Lev Ashlag (Sulam) and many others [1, 11, 25, 30].

Since its ancient days, Kabbalah proposed the Tree of Life as an integrative framework to understand the creation and development of the three basic levels of human existence: cognitive, emotional/behavioral and action levels. We will show how the Tree of Life contains elements for a system theoretic approach to economics and finance: feedback, hierarchical control etc. We will then show how to use these and the Tree of Life to integrate personal knowledge theory [34, 35], emotional intelligence [17–19, 38] with classical economic theory and the findings of behavioral finance and economics [2, 9, 13, 21, 22, 36, 39, 40].

In the second part we will develop for the first time system dynamics models, expanding further our earlier Kabbalah system theory to model knowledge based and behavioral economics and finance:

(1) Behavioral asset pricing system models including macro and micro economic factors, asset market technical and quantitative factors and asset prices together with an emotional and cognitive knowledge level
(2) Aggregate supply and demand behavioral and knowledge based models for goods prices
(3) Knowledge based production models linking invested capital, labor force and total production with knowledge and behavior (knowledge based and behavioral Cobb Douglas theory)
(4) Knowledge based and behavioral economic sectorial models linking productive sectors, consuming sectors, individual and household consumers, export (knowledge based and behavioral static and dynamic Leontieff sector balance models).

We will focus in particular on expanding (1) as an application example. Emotional and cognitive biases are formulated in this framework

- overconfidence bias: the excessive faith in one's cognitive abilities.
- conservatism bias: bias due to fixation in previous personal knowledge to the detriment of processing new economic and financial information (underreaction).
- optimism bias: an emotional bias, while the biases defined above were cognitive. It is the bias according to which investors, based on their emotional structure, are over-optimistic about markets and economical situation.

2 The Tree of Life of Kabbalah as the Framework for a Kabbalah System Theory

According to Kabbalah, human existence, the physical and psychological, emotional world and the process of its structuring and creation have ten fundamental general attributes/qualities called "sefirot", grouped in three categories [30]:

(1) knowledge and cognitive level (including objective and spiritual knowledge): Crown (will, faith and desire, Keter in original Hebrew or Aramaic), Wisdom (Chochmah), Understanding (Binah) and Knowledge (Da'at) which in fact prepares the transition and implementation of understanding at the emotional level. We are not going here into the detailed structure of this sefira, we did so in [6]
(2) emotional/behavioral level: Lovingkindness (Chesed), Judgment, Justice, Strength, Rigor or Severity (Gevurah) and Harmony or Beauty (Tiferet) which is connected to the next level below
(3) action level: Perseverance or Endurance (Netzach), Victory or Majesty (Hod), Foundation (Yesod) and Kingship (Malchut).

These ten fundamental attributes of the creation, development and existence processes are called in Hebrew sefirot (plural, sefira singular) which means counts, fundamental units. Despite their metaphorical anthropomorphic names, they do represent a very general metaphoric coordinate system of 10 general basic attributes (11 including Knowledge which normally is not represented in the same time with Crown), properties, actions that can be used to describe complex systems in general. In the Tree of Life, the ten sefirot fundamental units or components are interconnected by 22 arcs based on the interactions between them and between each of the three fundamental levels described above, in which these sefirot are integrated.

The internal sub-structure of each sefira is again of the type of a Tree of Life made of 10 sub-sefirot of the same type as the original 10 sefirot. This way, each sefira contains an internal model of the Tree of Life and of each of the sefirot it is in interaction with. In principle, we can go on and speak of the sub-sub-structure of sub-sefirot which will also be in the shape of Tree of Life etc. This means that the Tree of Life has a fractal structure or an inter-inclusive structure. However, for purposes of our Kabbalah system theory we will restrict ourselves to the first order sub-structure of the Tree of Life described by sub-sefirot of sefirot (see Fig. 1).

The names of the sefirot should be understood in their whole metaphoric symbolic generality. Lovingkindness for example, is the sefira of expressing emotions, of producing, of accepting. Judgment is the sefira of judging and understanding emotions, of consumption, of rigor, discipline, aversion and rejection.

The Tree of Life can also be seen as a system made up of three triadic levels. This is the simplified structural representation that we will use here though the missing sefirot can always be added to it:

(1) Cognitive: Wisdom-Understanding-Knowledge (ChabaD from Hebrew Chochmah-Binah-Dat denoted CBD),
(2) Emotional and Behavioral: Lovingkindness-Judgment-Harmony (ChaGaT from Hebrew Chesed-Gevurah-Tiferet denoted C'GT),
(3) Action: Endurance-Majesty-Foundation (NHY from Hebrew Netzach-Hod-Yesod denoted NHY).

Just like each sefira is made of 10 sub-sefirot, so each triad can be seen in its turn to be made up of three sub-triads. The Tree of Life has an inter-inclusive structure both in terms of sefirot and triads.

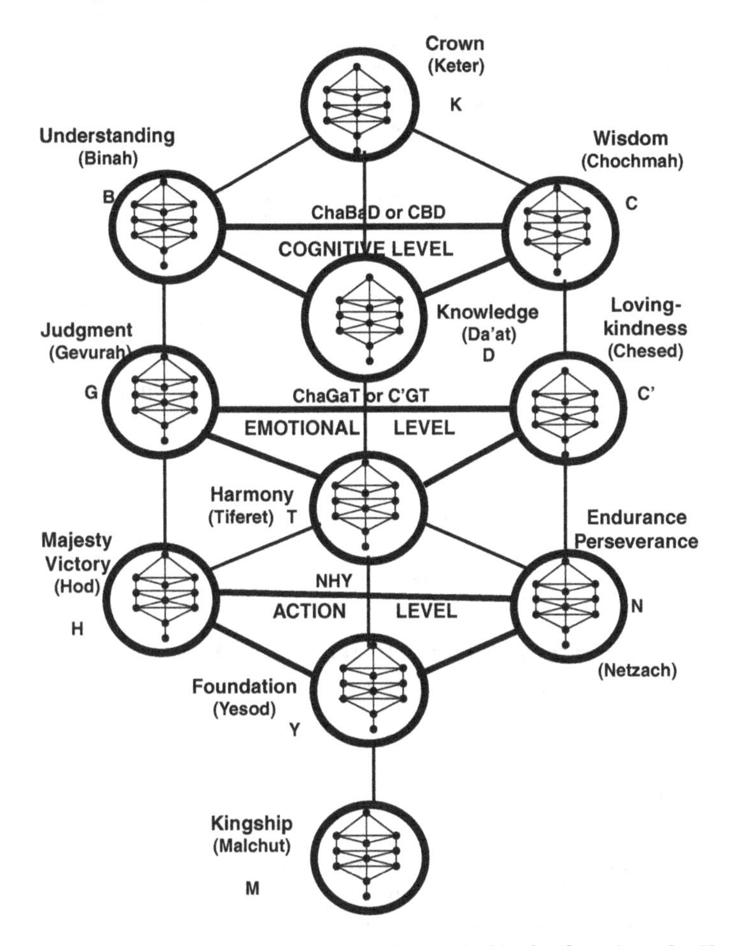

Fig. 1 The structure of the Tree of Life of Kabbalah model with 10 sefirot plus sefira Knowledge and the Cognitive (ChaBad or CBD), Emotional-Behavioral (ChaGaT or C'GT) and Action (NHY) triadic levels. Each sefira is described by its own similar sub-Tree of Life structure

The Tree of Life can also be seen in terms of the configurations, sets of sefirot (partzuf, singular, or partzufim, plural in Hebrew, Aramaic). For example, the five sephirot around Harmony plus sefira Harmony itself (Lovingkindness, Judgment, Endurance, Majesty, Foundation and Harmony itself) form the configuration of Harmony (Small configuration or Zeir Anpin in Hebrew, Aramaic).

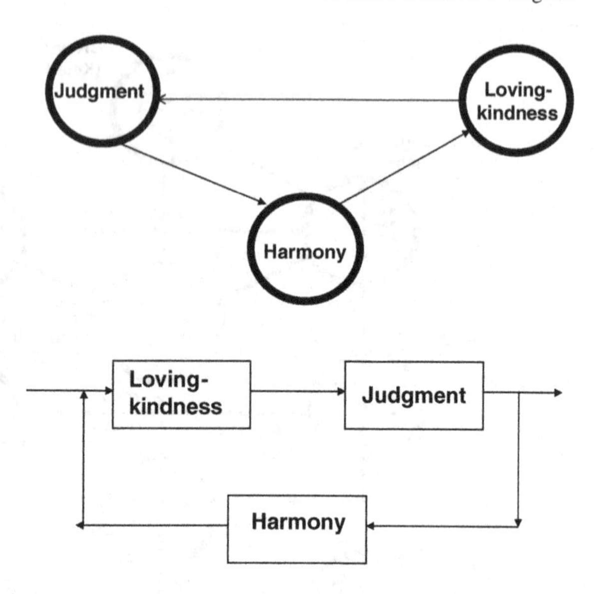

Fig. 2 The Tree of Life Emotional triad Lovingkindness-Judgment-Harmony (Chesed-Gevurah-Tiferet in Hebrew or ChaGaT denoted C'GT) as a feedback control system

3 A Kabbalah System Theory: Hierarchical Feedback Control Systems in the Tree of Life

The three sefirotic triads of the Tree of Life can each by modeled by a feedback control system as we showed in [6].

For example (see Fig. 2), the sefirotic triad made of Lovingkindness-Judgment-Harmony (abbreviated as ChaGaT or CGT from the initials Chesed-Gevurah-Tiferet of the corresponding Hebrew words), functions as an emotional level feedback control system with Harmony as feedback control helping Lovingkindness to regulate Judgment and viceversa. We have seen before how Lovingkindness and Judgment have opposing though complementary functions. Harmony is known in Kabbalah to represent the sefira of the middle equilibrium line that helps maintain the balance between Lovingkindness and Judgment and the intuition of feedback loops was there for a long time [3, 11]. In specific human and social system applications, Lovingkindness and Judgment can each be described by their own 10 sub-sefirot as we discussed earlier.

Based on the above, we proposed a three level hierarchical feedback control system model for the Tree of Life, made of hierarchically interconnected feedback control system models of knowledge-cognitive level, emotional level and action level (see Fig. 3).

In the Tree of Life of Kabbalah, flow of information goes both ways between sefirot. However, for simplicity, we represent in Fig. 3, only one sense of arrows in the CBD, C'GT, NHY levels compatible with the feedback control structure of these levels but one should bear in mind that flows of information at these levels are much more complex and can also go in the opposite sense to the one represented in Fig. 3.

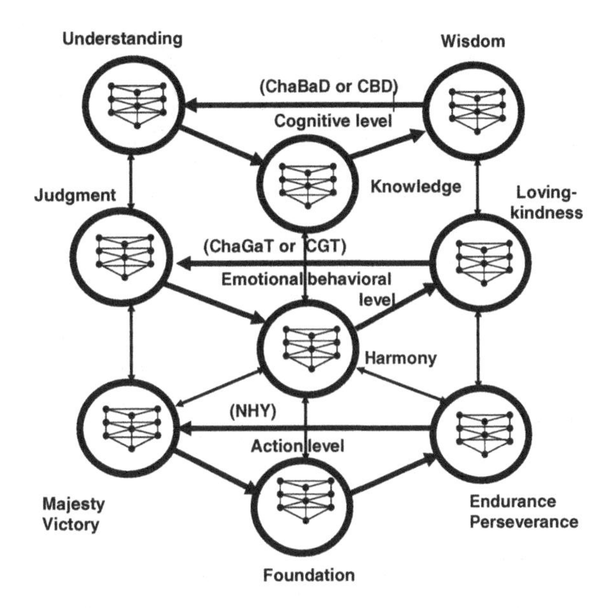

Fig. 3 The Tree of Life as a hierarchical three level feedback control system: Cognitive (Wisdom-Understanding-Knowledge (CBD)), Emotional-Behavioral (Lovingkindness-Judgment-Harmony (C'GT)) and Decision Making-Action (Endurance-Majesty or Victory-Foundation (NHY))

This hierarchical feedback control system in Fig. 3 is in fact a complex of horizontal and vertical, short and long feedback loops, descending and ascending, providing a range of feedback types in particular for complex economical and financial systems with behavioral and knowledge factors.

4 An Algebraic Combinatorial Modeling Framework for Kabbalah System Theory Based on Category Theory and Algebraic Topology

The Tree of Life with its 10 sefirot as nodes and its 22 directed edges or arcs connecting the sefirot defines a graph. What makes this graph special is the existence of three hierarchic levels in it corresponding to cognition, emotion and action, the symmetry left-right between the side of Lovingkindness and the side of Judgment with a middle axis of Harmony moderation between them. If we now consider the structure of each sefirot with its sub-Tree of Life, we do get a very specific Tree of Life graph indeed having an inter-inclusive structure, a graph of graphs, where the graph in each node has the same shape as the large overall graph. The arcs connecting two sefirot can be interpreted as transformations between their sub-structures modeled by graphs. This indicates that the Tree of Life, according to category theory [24, 26, 41], is in fact best described by a commutative diagram in the category of graphs.

The category of graphs is made of graphs as objects and graph transformations as morphisms including the possible compositions of such morphisms. Such graph morphisms or transformations between graphs are a model for the dynamic transitions

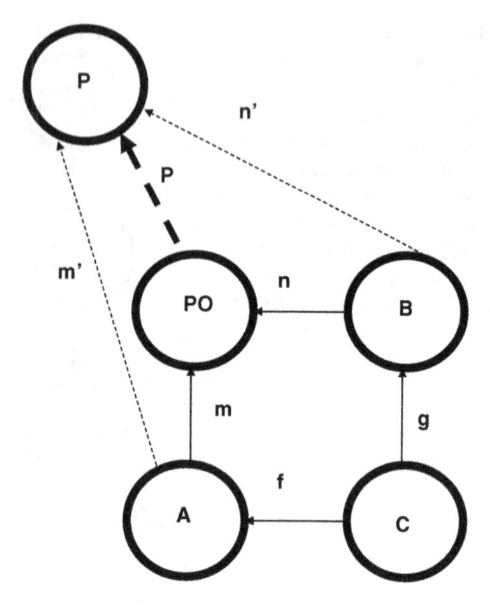

Fig. 4 The category theoretic commuting diagram definition of the pushout PO of objects A and B over C including the universality property (stability, robustness) of PO with respect to any other P

of a graph structure in time and hence for the dynamics of the Tree of Life graph. Graphs form a special type of category called "topos" [41] which has many useful properties for system theory modeling among which is the existence of pullbacks and pushouts which we define next for the case of a general category. For category theory and topos theory introduction see [24, 27].

Definition 1 The pushout of objects A and B over object C in a category containing objects A, B, C connected by the morphisms g: C → B , f: C → A, is an object PO of that category together with morphisms n: B → PO and m: A → PO in the category morphism set such that (i) the diagram in Fig. 4 commutes m o f = n o g (where "o" denotes morphism composition) and (ii) PO has the universality property meaning that for any other object P in the category and morphisms $m' : A \rightarrow P$ and $n' : B \rightarrow P$ that satisfy the commutativity of the diagram in Fig. 4 $m' o\ f = n'\ o\ g$, there exists a unique morphism p: PO → P such that $p\ o\ m = m'$ and $p\ o\ n = n'$ (see Fig. 4).

We introduce next a concept dual to pushout in category theory, pullback, obtained by reversing the morphism arrows in Definition 1.

Definition 2 The pullback of objects A and B over object C in a category containing objects A,B,C connected by the morphisms g: B → C , f: A → C, is an object PB of that category together with morphisms n: PB → B and m: PB → A such that (i) the diagram in Fig. 5 commutes f o m = g o n where "o" denotes morphism composition and (ii) PB has the universality property meaning that for any other object, P, in the category and morphisms $m' : P \rightarrow A$ and $n' : P \rightarrow B$ that satisfy the commutativity of the diagram in Fig. 5 that is $f\ o\ m' = g\ o\ n'$, there exists a unique morphism p: P → PB such that $m\ o\ p = m'$ and $n\ o\ p = n'$ (see Fig. 5).

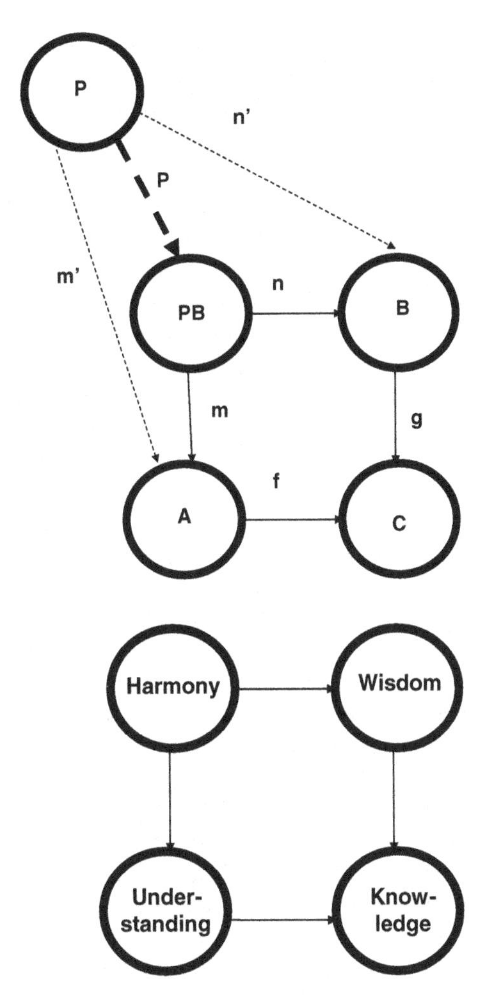

Fig. 5 The category theoretic commuting diagram definition of the pullback PB of A and B over C including the universality property (stability, robustness) of PB with respect to any other P

Fig. 6 In the Tree of Life, sefira Harmony is built through the pullback of wisdom and understanding over knowledge

Goguen [15, 16] showed how general system theory can be formulated within the algebraic framework of category theory and how pullback is a model for input-output behavior of systems while pushout is a model for system interconnections, couplings, feedbacks. Negoita [32, 33] and Negoita and Ralescu [31], based on Goguen, introduced pullback in expert systems and human system management models.

For example, in Burstein and Negoita [6] it is shown in great detail how in the Kabbalah of Tree of Life, Harmony is built through a pullback of Understanding and Wisdom over Knowledge (see Fig. 6). To be precise, according to Kabbalah, the cognitive subsystem CBD (brain or mochin in Hebrew) of Configuration of sefirot around Harmony, Zeir Anpin (ZA), is created as a pullback of Wisdom CBD and Understanding CBD over Knowledge.

We will generalize this further here to involve more sefirot at all levels and multiple limits or nested limits (limits of limits).

Pullback and pushout are two particular cases of the more general concepts of limits and colimits in a category [26, pp. 62–72]. These are defined similarly as in Figs. 4 and 5 except that instead of two objects A, B mapped into a third object C, an arbitrarily complex commuting diagram **D** of objects and morphisms between them is used instead. The limit for example, the multidimensional generalization of pullback, is given by an object and morphisms from it to each of the objects of the diagram **D** such that the overall diagram commutes and the limit object has universality property defined as a multidimensional generalization of the universality property in Definition 2 (we say that we have a limit or limiting cone for diagram **D**). Similarly we have colimits or limiting cocones..

The Tree of Life has more than just nodes and arcs, it also has triads or triangles as higher dimensional faces like CBD, C'GT, NHY and each sefira has its own sub-triads or sub-triangles as shown in Fig. 1. If we ignore for the moment the sub-Trees of Life that make up each individual sefirot, the Tree of Life can be seen as an abstract simplicial complex, a concept from algebraic topology and its category theory formulation [14, 29] which we introduced in complex dynamic system theory modeling [5, 7, 8]. Thel Tree of Life is made of 0-dimensional simplices, the sefirot, 1-dimensional simplices or faces, the arcs, and 2-dimensional simplices or faces, the triads or triangles. An abstract simplicial complex is a multidimensional generalization of graphs as it allows faces (simplices) of higher dimensions.

Considering each sefirot modeled by its own simplicial complex structure as above, the Tree of Life becomes a commutative diagram in the category of abstract simplicial complexes. This is not a topos as was the category of graphs but, nevertheless, has finite limits and colimits, pullbacks and pushouts [20]. In this category, abstract simplicial complexes are the objects and simplicial maps between simplicial complexes are the morphisms. Simplicial maps map faces into faces or simplices into simplices. Simplicial maps can model dynamic transitions in The Tree of Life.

In Burstein and Nicu [7] and Burstein et al. [8] a simplicial dynamic system theory was introduced on abstract simplicial complexes (such as the Simplicial Tree of Life in Fig. 1), using multidimensional simplicial vector fields and their flows on simplicial complexes.

We have now all ingredients to assemble the category theoretic and algebraic topological combinatorial model of the Tree of Life to be used in our Kabbalah system theory (see Fig. 7).

Each sefira is represented by its 2-dimensional abstract simplicial complex (sub-Simplicial Tree of Life model) as in Fig. 3 (2-dimensional because we consider each sefira as having its own CBD, C'GT, NHY triads as 2-dimensional simplices). We denote these abstract simplicial complexes by K(C), K(B), K(D)... in Fig. 7 for each of the sefira C, B, D,.... Simplicial maps between the simplicial complexes of sefirot, mapping simplices of one sefira into simplices, of the other sefira are used in Fig. 7 as algebraic model for the connections between sefirot in terms of their constituent sub-simplices. We denoted these simplicial maps by F(CB), F(BD), F(CD)...corresponding to the sefirot which are linked.

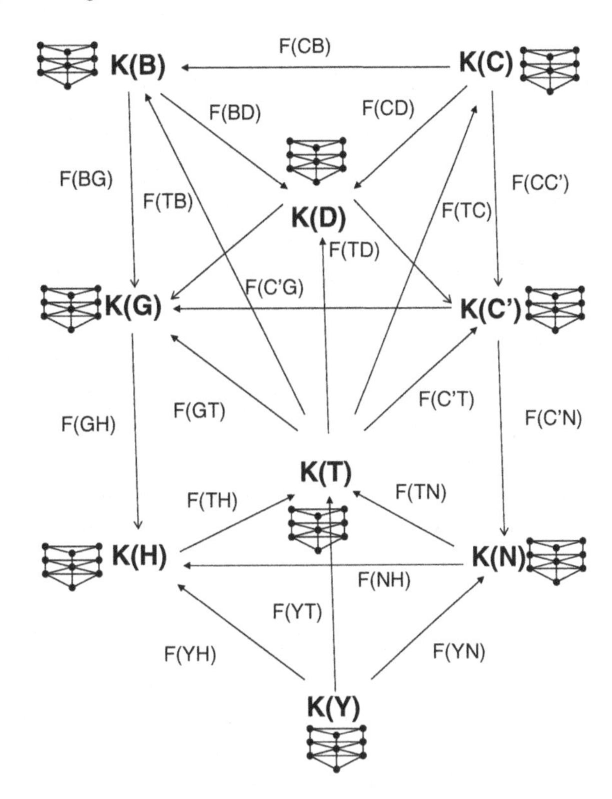

Fig. 7 The category theoretic and algebraic topological combinatorial modeling framework for the Tree of Life hierarchic system in Fig. 3: each sefira is represented by an abstract simplicial complex K(C), K(B), K(D)... and sefirot are connected by simplicial maps that map simplices into simplices, F(CB), F(CD), F(BD).... The overall diagram must commute and so does any sub-diagram, as diagrams in the category of abstract simplicial complexes. All limits and co-limits (pullbacks and pushouts) of any sub-diagram do exist in the category of simplicial complexes

The diagram in Fig. 7 with simplicial complexes and simplicial maps in the category of simplicial complexes must be commutative including any sub-diagrams of it. All the limits and colimits (pullbacks and pushouts) of sub-diagrams in Fig. 7, do exist indeed in the category of simplicial complexes [20]. In the Tree of Life, as we said, the flow of information goes both ways between two sefirot which amounts in fact to the existence of two arrows both ways between each two sefirot in Fig. 7. However, for simplicity, we only represent the arrows that will be used in the behavioral and knowledge based economics and finance modelling and focus on Limits but bear in mind that dual arrows exist too and so do colimits of these.

In addition, we can assume K(T) to be equal to the limit (pullback multidimensional generalization) of the diagram (Lim (K(C), K(B), K(D)), K(C′), K(G)) which exists in the category of simplicial complexes and where Lim (K(C), K(B), K(D)) is the limit of diagram (K(C), K(B), K(D)) which exists. Given the universality property of Lim (K(C), K(B), K(D)), there exists a morphism from K(T) to this limit and hence Lim (Lim(K(C), K(B), K(D)), K(C′), K(G)) is well defined in the sense that K(T) can form indeed a limiting cone for (Lim (K(C), K(B), K(D)), K(C′), K(G)). The assumption on K(T) can be considerably relaxed using the universality property of Lim (Lim(K(C), K(B), K(D)), K(C′), K(G)) which implies the existence of a morphism from K(T) to Lim(Lim (K(C), K(B), K(D)), K(C′), K(G)) which exists.

These assumptions on K(T) generalize the construction of K(T) as pullback of K(C) and K(B) over K(D) detailed in [6] or the construction of K(T) as limit of the commuting diagram (K(C), K(B), K(D)). All the above limits exist in the category of simplicial complexes subject to the commuting diagram in Fig. 7 and they are a model of system input-output behavior [15, 16]. We also assume K(Y) to be the limit of (K(T), K(N), K(H)) commuting diagram and thus being equal to its input-output behavior. Alternatively, we can assume K(Y) to be equal to the pullback of K(H) and K(N) over K(T). Again, all the above assumptions and construction formulas can be relaxed by using the universality property of limits which yields for example a morphism from K(Y) to Lim (K(N), K(T), K(H)) which exists.

The limit assumption K(T) = Lim (Lim (K(C), K(B), K(D)), K(C′), K(G)) means that the behavior of the Knowledge level control system (K(C), K(B), K(D) can be calculated within the Tree of Life and is learned by K(T) which internalizes Knowledge at the Emotional level aggregating it with emotional K(G), K(C′) in order to transmit this to Action level. Pullback and limits amount to aggregating and synthesizing behavior of systems at different levels.

5 System Dynamics in Kabbalah System Theory

We will expand here the Kabbalah system theory framework created in [6] by developing a system dynamics model for the Tree of Life. In a category theory framework, the simplest model of discrete time dynamics for an object is an endomap or endomorphism from the object to itself [24, p. 137]. We consider simplicial endomaps d(B), d(C), d(C')... for each of the simplicial complexes K(B), K(C), K(C')... in Fig. 7, mapping each of the simplicial complexes to itself in the category of simplicial complexes through simplicial maps. This leads to 3 groups of dynamics:

KNOWLEDGE DYNAMICS	KD = <d(C), d(B), d(D)>
EMOTIONAL DYNAMICS	ED = <d(C'), d(G), d(T)>
ACTION DYNAMICS	A = <d(N), d(H), d(Y)>

In order to get a model for the dynamics of the Tree of Life algebraic combinatorial model in Fig. 7 we impose that the diagram in Fig. 8 commutes meaning that this is also the case for all sub-diagrams in the category of simplicial complexes that can be made with any of the complexes K(B), K(C), K(C')..., any of the endomaps d(C), d(B), d(D)...of these complexes into themselves and any of the simplicial maps F(CB), F(CD), F(BD)...from the Tree of Life model in Fig. 7 (see Fig. 8). Figure 8 basically says that the Tree of Life connects Knowledge Dynamics, Emotional Dynamics and Action Dynamics between themselves in a way that is compatible with and influenced by the relation between sefirot in the Tree of Life. The dynamics of the Tree of Life is ultimately given by the three dynamics.

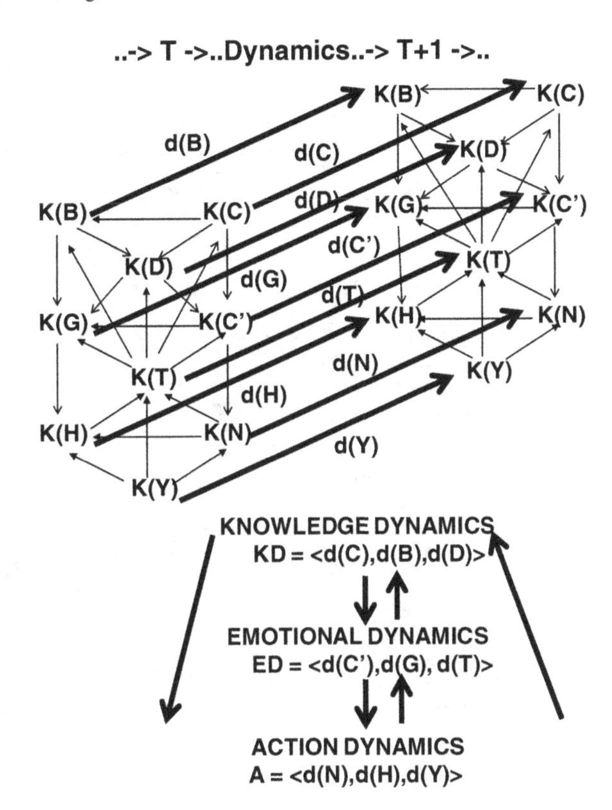

Fig. 8 System dynamics model in the Kabbalah system theory of the Tree of Life given by a commutative diagram in the category of simplicial complexes. Simplicial endomaps or endomorphisms between each sefira simplicial complex and itself describe local dynamics at the level of each sefira. Commutativity of the overall diagram means that sefirot dynamics are compatible with and influenced by the relation between sefirot in the Tree of Life. The overall dynamics maps the Tree of Life algebraic combinatorial model in Fig. 7 into itself

6 The Kabbalah System Theory Modeling of Behavioral and Knowledge Based Economics and Finance

We can finally apply the Kabbalah system theory modeling framework, developed in the previous sections, to economics and finance. A hierarchical feedback control Tree of Life model like the one in Fig. 3 will be used to integrate together personal knowledge, emotional intelligence elements and elements of economics and finance theory (see Fig. 9).

The Cognitive, knowledge level of the economic and financial decision making process, can be adequately represented in terms of Polanyi personal or aggregated knowledge theory as a framework for dynamic knowledge [34, 35]. Tacit knowledge is the difficult to articulate, subconscious/unconscious knowledge based on experience [28]. Explicit knowledge is decoded, explained tacit knowledge, formalized, documented, organized, conscious etc. Once tacit knowledge becomes explicit knowledge, it is categorized or classified according to action and behavioral emotional frames so that it can be used by the next levels, the level of emotions and behavior and by the level of actions. Popper's objective knowledge theory also has three levels called "three worlds" and can be alternatively used here [37].

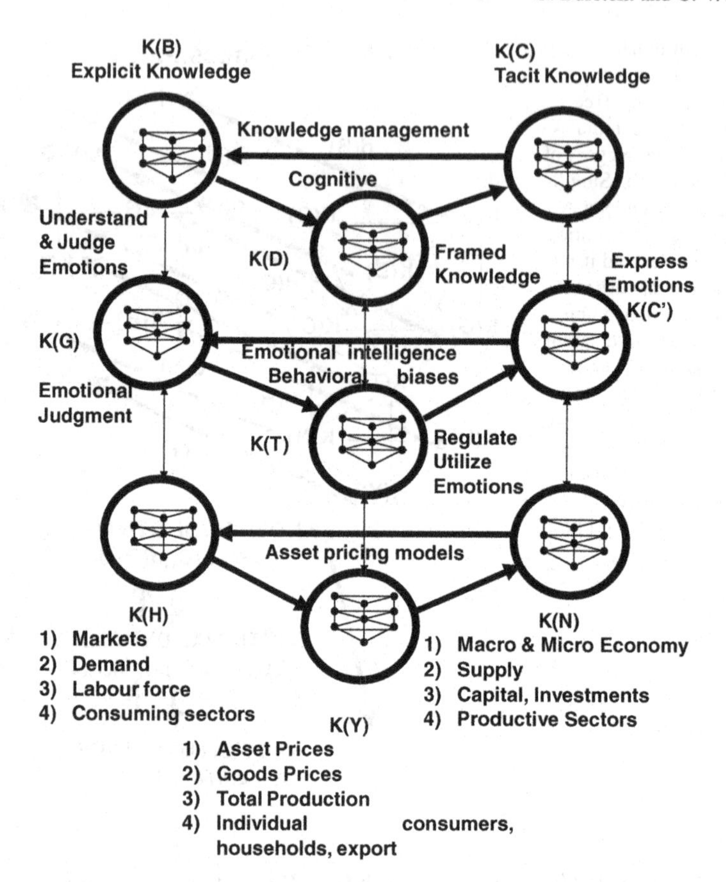

Fig. 9 Kabbalah system theory modeling for (1) behavioral and knowledge based asset pricing, (2) aggregate supply and demand behavioral and knowledge based model for goods prices, (3) knowledge based production models of Cobb-Douglas type linking invested capital, labor force and total production with knowledge and behavior and (4) knowledge based and behavioral economic sectorial balance models of Leontieff type linking productive sectors, consuming sectors, individual and household consumers and export

Emotional intelligence theory [17–19, 38] offers the ideal way to model the Emotions behavioral level in the Tree of Life structure in Fig. 8. The three components of the triad are expressing emotions, understanding and judging emotions and regulate/control and utilize them for action level.

The bottom triad of Action is the actual classical economics and finance processes level. Finance and classical economic theory [2, 13] can be used to obtain Tree of Life system models according to Fig. 9 for:

(1) Behavioral and knowledge based asset pricing system modeling including macro and micro economic factors, asset market technical and quant factors and asset prices together with emotional and knowledge factors (select items labeled with (1) for each of the sefirot N, H, Y)

(2) Aggregate supply and demand behavioral and knowledge based modeling for goods prices (by selecting items labeled with (2) for each of the sefirot N, H, Y)
(3) Knowledge based production models linking invested capital, labor force and total production with knowledge and behavior giving knowledge based and behavioral Cobb Douglas theory for example (by selecting items labeled (3) for each of the sefirot N, H, Y)
(4) Knowledge based and behavioral economic sectorial models linking productive sectors, consuming sectors, individual and household consumers, export giving knowledge based and behavioral static and dynamic Leontieff sector balance models (by selecting items (4) for each of the sefirot N, H, Y).

This way, Fig. 9 displays together four possible behavioral and knowledge based models for four categories of economic or financial topics within the Tree of Life hierarchical feedback control systems as in Fig. 3.

Although supply, demand, consumption, investments, production seem clearly objective economic variables, they are all subjected to behavioral elements such as fear. We already learned after September 11, 2001 and after the 2008 financial crisis, that consumer fears and investor fears can trigger behaviorally induced economic slowdowns despite any objective signs of early economic recovery.

Kabbalah system theory expanded in the previous section (see Figs. 7 and 8) can be used to create the mathematical computational models for Fig. 9 according to the commuting diagram with limits and colimits (pullbacks and pushouts) in Figs. 7 and 8.

In this framework, we can approach the knowledge sensitivity of demand and goods prices pioneered in [28] based on Polanyi personal knowledge theory [34, 35]. We consider choice of items (2) in Fig. 9: Supply-Demand-Goods Prices. The knowledge based dynamics of supply and demand affecting prices and vice-versa via feedbacks, can be described by the corresponding model in Figs. 7 and 8 for the Tree of Life system model in Fig. 9, choice (2). Based on the assumptions/constructions on K(Y) and K(T) made in the discussion after formulating the commuting diagram model in Fig. 7, goods prices appear as a multiple limit (multidimensional generalization of pullback) involving Supply, Demand, Explicit and Tacit knowledge, Framed knowledge, and the expressed and judged/understood emotions given by K(C') and K(G) as follows:

Goods Prices = Lim (Supply, Demand, Lim (Lim (Tacit K, Explicit K, Framed K), Express Emotions, Judge/Understand Emotions)).

As we explained above, limits of diagrams in categories represent input output behavior of systems described by those diagrams [16]. Calculating the above limit opens the door to simulating input-output behavior of the knowledge based behavior of goods prices.

Given that K(T) in choice (2) of Fig. 9, represents regulating and using emotions in decision making and actions of supply and demand processes, the K(T) limit assumption/construction over knowledge level limit and over expressed and understood/judged emotions, shows algebraically how emotional intelligence, the regulator of emotions is using decoded and emotionally categorized personal knowledge.

Cobb-Douglas production functions and Leontieff sector balance models can be formulated in Fig. 8 with choices (3) and respectively (4) from Fig. 9. Total production appears as limit of Labor force, capital and investments but also explicit, tacit and framed knowledge filtered through our emotional intelligence.

Behavioral finance cognitive and emotional biases and heuristics [36, 40] create market mispricings at stock price and stockmarket index levels. Behaviorally biased interpretation of company microeconomic factors (earnings results) and stock price moves is not the only source of mispricings, there is also macroeconomic information which is behaviorally interpreted causing macroeconomic mispricings such as those arbitraged by the global macroeconomic arbitrage introduced in [4] and analysed by Werner de Bondt, one of the pioneers of behavioral finance [12]. Using choice (1) in Fig. 9 and the corresponding category theoretic simplicial complex system dynamics given by the commuting diagram in Fig. 8 for the complexes in Fig. 9, we obtain the dynamics for behavioral and knowledge based asset pricing (BKAP):

KNOWLEDGE DYNAMICS $\quad KD = <d(C), d(B), d(D)>$

EMOTIONAL DYNAMICS $\quad ED = <d(C'), d(G), d(T)>$

BEHAVIORAL AND KNOWLEDGE
BASED ASSET PRICING DYNAMICS $\quad BKAP = <d(N), d(H), d(Y)>$

Based on Fig. 9 we have the following Kabbalah behavioral, knowledge based asset pricing model:

$$\text{Asset prices } K(Y) = \text{Lim}\ (K(H), K(N), K(T)) =$$

$$= \text{Lim}\ (K(H), K(N), \text{Lim}\ (\text{Lim}\ (K(B), K(C), K(D)), K(C'), K(G))$$

where we used in the above formulas

$$K(T) = \text{Lim}\ (\text{Lim}\ (K(B), K(C), K(D)), K(C'), K(G))$$

as discussed after introducing the commuting diagram model in Fig. 7.

This model is static. In order to get the dynamic model, we replace all simplicial complexes by their corresponding "simplicial complex in time" presheaves and morphisms of presheaves of simplicial complexes in functor categories or categories of presheaves (see [27, pp. 25, 36] for presheaves in time concept). We get commuting diagram of presheaves of simplicial complexes in time in Fig. 10.

Alternatively we can use the dynamics $ED = <d(C'), d(G), d(T)>$ and $BKAP = <d(N), d(H), d(Y)>$ defined in Fig. 8 in order to propagate and map in time $K(Y) = \text{Lim}\ (K(H), K(N), K(T))$ and $K(T) = \text{Lim}\ (\text{Lim}\ (K(B), K(C), K(D)), K(C'), K(G))$ through the commuting diagram in Fig. 8.

In this framework we can model the dynamics of emotional and cognitive biases and heuristics influence on asset prices:

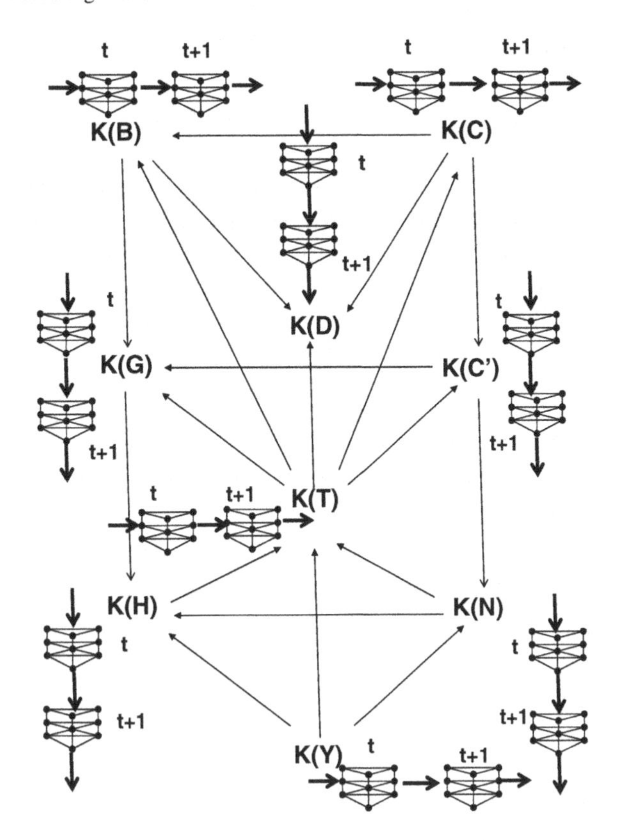

Fig. 10 Kabbalah system theory model for behavioral and knowledge based asset dynamics of the Tree of Life in Fig. 9, case (1). This is a commutative diagram in the category of presheaves of simplicial complexes in time. Each sefira is described by a presheaf of simplicial complexes in time and the maps between sefirot are morphisms of presheaves of simplicial complexes in time. Commutativity of the overall diagram means that sefirot dynamics are compatible with and influenced by the relation between sefirot in the Tree of Life

- overconfidence bias is the excessive faith in one's cognitive abilities. This is modeled within the framework of Fig. 7 described in the previous section: K(Y), representing asset prices, is modeled as a limit or limiting cone (multidimensional generalization of pullback) of K(H), K(N), K(T) commuting diagram representing respectively markets, macroeconomic and microeconomic company factors and explicit K(B) and tacit K(C) knowledge filtered through our emotional intelligence
- conservatism bias is due to fixation in previous personal knowledge to the detriment of processing and acknowledging new economic, markets and company information. This leads to prices in K(Y) to underreact to new market and economic information in K(H) and K(N) and rather to price in excessively previous explicit knowledge K(B) framed as K(D)
- optimism bias is an emotional bias, while the biases defined above were cognitive. It is the bias according to which investors, based on their emotional structure, are over-optimistic about markets and economic situation. This can be explained in our model of Fig. 7 by the algebraically aggregated effect of the emotional judgment (K(G)) component of emotional intelligence and the emotionally categorized explicit knowledge in K(D).

Conclusions

We have developed here a Kabbalah system theory approach to behavioral and knowledge based economics and finance. This allowed us to integrate elements of classical economic and finance theory with Kabbalah, personal knowledge theory and emotional intelligence theory to create a modeling framework.

The Tree of Life of Kabbalah is ideal for this framework since it has three interconnected levels: cognitive, behavioral and action. We developed here system structure and dynamics models for the Kabbalah system theory of the Tree of Life reflected in Figs. 7, 8, 9 and 10 based on category theory and algebraic topology.

Pullback and pushout (limit and colimit) in category theory were used to model system behavior and respectively, system interconnection.

We focused particularly on formulating behavioral knowledge based asset pricing models and cognitive and emotional biases in this Kabbalah system theory modeling framework.

Acknowledgments The first author would like to express his gratitude for great teachings and guidance to R. Dangur, A. Babayove, M. Krantz, M. Handler, M. Millstein, N. Citron, S. Anteby, N. Blavin, A. Sutton, A. Greenbaum, R. Afilalo, Y. Ginsburgh. We are also very grateful to the anonymous referee for many suggestions including to develop further system dynamics models for Kabbalah system theory.

References

1. Afilalo, R.: The Kabbalah of the Arizal According to the Ramhal. Kabbalah Editions, Montreal (2004)
2. Begg, D., Fischer, S., Dornbusch, R.: Economics. McGraw Hill Education, Berkshire (2005)
3. Bernard-Weil, E.: Du Systeme a la Torah (French Edition). L'Harmattan, Paris (1995)
4. Burstein, G.: Macro Trading and Investment Strategies: Macroeconomic Arbitrage in Global Markets. Wiley, New York (1999)
5. Burstein, G., Ivascu, D.: An exact sequences approach to the controllability of systems. J. Math. Anal. Appl. **106**(1), 171–179 (1985)
6. Burstein, G., Negoita, C.V.: Foundations of a postmodern cybernetics based on Kabbalah. Kybernetes **40**(9–10), 1331–1353 (2011)
7. Burstein, G., Nicu, M.D.: Simplicial differential geometric hierarchical control of large scale systems. In: Geering, H.P., Mansour, M. (eds.) Large Scale Systems: Theory and Applications, pp. 464–468. Pergamon Press, Oxford (1986)
8. Burstein, G., Nicu, M.D., Balaceanu, C.: Simplicial differential geometric theory for language cortical dynamics. Fuzzy Sets Syst. **23**, 303–313 (1987)
9. Camerer, C.F., Loewenstein, G., Rabin, M. (eds.): Behavioral Economics. Russell Sage Foundation and Princeton University Press, New York (2004)
10. Collins, H.: Tacit and Explicit Knowledge. University of Chicago Press, Chicago (2010)
11. Cordovero, M.: Pardes Rimonim: Orchard of Pomegranates. Providence, Belize City (2010)
12. de Bondt, W.: Global macro-economic arbitrage: the investment value of high-precision forecasts. Revue de la Banque/Bank- en Financiewezen (2005)
13. Dornbusch, R., Fischer, S., Startz, R.: Macroeconomics. McGraw Hill/Irwin, New York (2007)
14. Giblin, P.J.: Graphs, Surfaces and Homology. Cambridge University Press, Cambridge (2010)

15. Goguen, J.: Mathematical representation of hierarchically organized systems. In: Attinger, E. (ed.) Global Systems Dynamics, pp. 112–128. Wiley Interscience, New York (1970)
16. Goguen, J.: Sheaf semantics for concurrent interacting objects. Math. Struct. Comput. Sci. **11**, 159–191 (1992)
17. Goleman, D.: Emotional Intelligence. Bantam Books, New York (2005)
18. Goleman, D.: Working with Emotional Intelligence. Bantam Books, New York (2006)
19. Goleman, D.: Social Intelligence. Bantam Books, New York (2007)
20. Grandis, M.: Higher fundamental functors for simplicial sets. Cah. Topol. Geom. Diff. Catégoriques **42**, 101–136 (2001)
21. Kahneman, D., Slovic, P., Tversky, A. (eds.): Judgment under Uncertainty: Heuristics and Biases. Cambridge University Press, Cambridge (1998)
22. Kahneman, D., Tversky, A. (eds.): Choices, Values and Frames. Cambridge University Press and Russell Sage Foundation, Cambridge (2000)
23. Lange, O.: Introduction to Economic Cybernetics. Pergamon Press, London (1970)
24. Lawvere, F.W., Schanuel, S.H.: Conceptual Mathematics: A First Introduction to Categories. Cambridge University Press, Cambridge (2005)
25. Luzzato, M.C.: 138 Openings of Wisdom by Rabbi Moshe Chaim Luzzatto ("Ramchal"), translated by A.Y. Greenbaum. The Azamra Institute, Jerusalem (2005)
26. Mac Lane, S.: Categories for the Working Mathematician. Springer, New York (1998)
27. Mac Lane, S., Moerdijk, I.: Sheaves in Geometry and Logic: A First Introduction to Topos Theory. Springer, New York (1992)
28. Maracine, V., Ianole, R.: How knowledge determines demand dynamics. Econ. Comput. Econ. Cybern. Stud. Res. **44**(4), 5–22 (2010)
29. Maunder, C.R.F.: Algebraic Topology. Dover Publications, Mineola (1980)
30. Menzi, D.W., Padeh, Z.: The Tree of Life: Chayyim Vital's Introduction to the Kabbalah of Isaac Luria. Arizal Publications Inc., New York (2008)
31. Negoita, C.V., Ralescu, D.A.: Simulation, Knowledge Based Computing and Fuzzy Statistics. Van Nostrand, New York (1987)
32. Negoita, C.V.: Fuzzy Systems. Abacus Press, Tunbridge Wells (1981)
33. Negoita, C.V.: Expert Systems and Fuzzy Systems. Benjamin Cummings Pub. Co., Menlo Park (1985)
34. Polanyi, M.: Personal Knowledge. University of Chicago Press, Chicago (1962)
35. Polanyi, M.: Knowing and Being. University of Chicago Press, Chicago (1969)
36. Pompian, M.M.: Behavioral Finance and Wealth Management. Wiley, New York (2006)
37. Popper, K.R.: Objective Knowledge. Oxford University Press, Oxford (1979)
38. Salovey, P., Mayer, J.D.: Emotional intelligence. Imagination Cogn. Pers. **9**(1990), 185–211 (1990)
39. Shefrin, H.: A Behavioral Approach to Asset Pricing. Elsevier Academic Press, Burlington (2005)
40. Thaler, R.H. (ed.): Advances in Behavioral Finance, vol. II. Princeton University Press, Princeton (2005)
41. Vigna, S.: A guided tour in the topos of graphs, Technical report 199-97. Universita di Milano, Dipartimento di Scienze dell'Informazione (1997)
42. von Bertalanffy, L.: General System Theory. George Braziller, New York (1969)
43. von Bertalanffy, L.: Perspectives on General System Theory. George Braziller, New York (1975)
44. Wiener, N.: Cybernetics or Control and Communication in the Animal and the Machine. MIT Press, Cambridge, Massachusetts (1948)

Chapter 3
The Commodity Exporting Country A Spectral Analysis of Brazilian and Australian Equity Markets

David Gray and John McManus

Abstract Although the equity markets of the developed and developing world have been plagued by the vagaries of financial flows and the threat of contagion since the collapse of Lehman Brothers, the commodity-based economies of Brazil and Australia exhibit limited evidence of this. Rather than disrupting relations, financial information appears to be absorbed by equity market in an efficient manner, reflecting time-zone delays.

1 Introduction

Brazilian Finance Minister, Guido Mantega, proclaimed in September 2010 that the world is "in the midst of an international currency war" [15]. Currency appreciation, strong growth and high interest rates encourage speculators to invest in equity and bond investments in economies like Australia and Brazil. Guido Mantega is pointing the finger of blame for the resultant asset price inflation at the European Central Bank, the FED, the Bank of England and the Bank of Japan and extremely loose monetary policy.

There have been numerous studies of both currency and equity markets featuring the Asian currency crises (e.g. [7]); the October 1987 crash (e.g. [13]); and the Lehman Bros' 2008 banking crisis (e.g. [2]). Many papers have considered contagion. Contagion could be said to exist if, following an event or shock, there is a change in the degree of concordance among asset markets. Altering the correlations between

D. Gray
Lincoln Business School, University of Lincoln, Brayfordpool Campus,
Lincoln LN6 7TS, UK

J. McManus (✉)
Faculty Business, Education and Law, Staffordshire University, Brindley Building,
Leek Road, Stoke-on-Trent , StaffordshireST4 2DF, UK
e-mail: john.mcmanus@staffs.ac.uk

V. K. Mago and V. Dabbaghian (eds.), *Computational Models of Complex Systems*,
Intelligent Systems Reference Library 53, DOI: 10.1007/978-3-319-01285-8_3,

© Springer International Publishing Switzerland 2014

some of the constituents could undermine the efficacy of the portfolio of assets. Contagion implies the existence of a tranquil, pre-event period, which, following the event, acts as a benchmark for gauging change. Forbes and Rigobon [6] draw a distinction between interdependence and contagion. Contagion must be beyond normal market interdependence: there needs to be significant 'intensification' of the link.

Although they are thousands of miles apart and are at different stages of development, Brazil and Australia have key features in common. They are commodity exporters that are affected by the weather systems, El Niño and La Niña. Unusually for large economies, these two are both dependent on a third. During the current financial crisis, China appeared to power on, growing by eight percent a year whilst others struggled to avoid recession. China's veracious growth is fed by *inter alia* the commodity-abundant economies of Brazil and Australia. In 2010, Australia had a trade to GDP ratio of 44 % with China as its most important trading partner. It accounts for the largest share of its exports (25 %) and imports (18 %). Brazil is less Sino-focussed with the corresponding figures of a 23.8 % trading ratio and 16 % and 14 % for exports and imports, respectively. The structure of exports indicated is also different. Table 1 indicates that over 60 % of Australian exports by value are fuel and mining products, when only 28 % of Brazil's are these commodities.

Figure 1 presents monthly Iron Ore, Brent crude oil, Australian thermal coal prices and Commodity food and beverage Index data from the IMF over 4 years to 2010. Iron ore has a general upward trend whereas the other three show a commodity price collapse beginning in June/July 2008 and ending in the first quarter of 2009.

2 Equity Market Concordance

One measure of market integration is common price index movements. Export base models would predict that economic cycles of both Australia and Brazil would be influenced by the prices of the commodities they export. Beyond this, the extent to which equity markets are integrated would also affect by cross-country horizontal mergers, common ownership and cross-market share listings. The desire of global exporters to achieve greater economies of scale or to pursue a consolidation of producers would lead to both real and financial links. Although it stalled, one can see combining operations of extractors Rio Tinto and BHP Billiton in this light.

Table 1 Proportion of commodities in exports

	Australia exports (%)	Brazil exports (%)
Agricultural products	12.7	34.0
Fuels and mining products	61.1	27.9

Source WTO, 2010

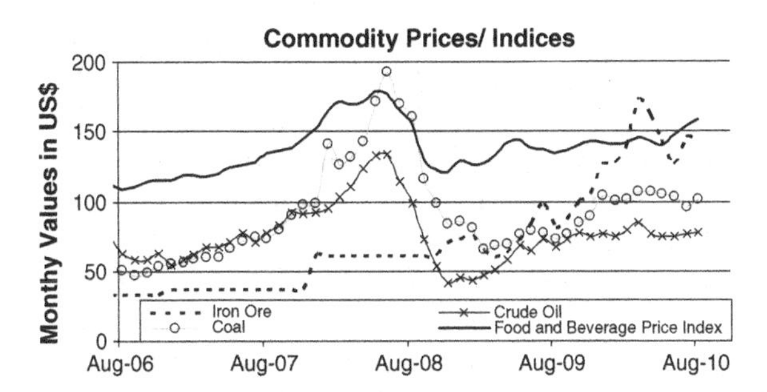

Fig. 1 Commodity Prices/Indices

Having similar macroeconomic fundamentals that run at odds with the US, Europe and Japan put Australia and Brazil as bulwarks against the poor outlook in more traditional equity markets. Their tight monetary stances were at odds with Quantitative Easing with which the US et al. were dabbling. The speculative investments that entailed significant capital inflows into Australia and Brazil could stimulate greater integration of their equity markets [7].

The dissimilarities may outweigh the above. Brazil has a weak record of financial rectitude, which could breed more turbulence. Capital mobility is imperfect, not least because Brazil has been a leading user of Tobin taxes. The wealth and the stage of development place Australia in a post-industrial era whereas Brazil is industrialising.

2.1 Considerations

There is a consideration of the degree of concordance between the São Paulo and Australian Stock Exchanges over four years. The thesis is that a contagion inducing shock occurred when Lehman Brothers collapsed (15th September, 2008) provoking equity market turbulence. In addition, loose monetary policies elsewhere contribute further volatility. Subjected to capital inflows, the Australian and Brazilian equity markets may become more integrated, increasing the degree of concordance.

Both the shock and the QE capital flows could have temporary effects, in which case, co-movements could return to the tranquil levels, or there is a permanent increase in the degree of integration. Turbulence and speculation are presumed to affect the higher frequencies disproportionally.

The work covers the two years before and two years after Lehman Brothers collapsed. The years before the collapse are pre-event period, which acts as a benchmark for gauging change the turbulent era. The year before the Lehman collapse could be used as the reference year. However, this year 1 is characterised by a banking crisis, with a sudden halt to the normal workings of the London Interbank Lending system, resulting in the failure of Northern Rock in the UK in September 2007, so an earlier year (September, 2006 to August 2007) is chosen as the tranquil era. The concordance between the currency-adjusted indices will be explored using spectral analysis. Greater co-dynamics will be reflected in an increase in cospectra. The extent to which there is intensification of concordance is assessed through change in coherence between the equity market indices. There is a review using Smith's [13] non-parametric approach to changes in coherence. Also, the power spectra and the error spectra are estimated to reveal whether distributions of [unexplained] perturbations become more similar; indicators of more similar price drivers.

The work is compiled as follows. In Sect. 2, there is a review of models of stock exchange interaction. The review of literature, in Sect. 3, concerns financial contagion over numerous crises and methods to establish its veracity. Sect. 4 offers a backdrop for the results. Sections 5 and 6 review the methods and data. It is shown in Sect. 7 that there has not been an intensification of co-movement between the São Paulo and Australian Stock Exchange. Despite capital movement and developmental, differences the relationships between the equity markets seem stable. Once a time-zone adjustment is made, the markets behave as if information is absorbed efficiently.

3 Literature Review

The 'traditional' approach to the relationship between the currency and equity markets of a sovereign state proposes that the two are linked through trade, and so emphasises the current account. The stock exchange will reflect the returns from investing in large, internationally oriented, firms. Export-oriented multinational firms benefit from a fall and importing ones, from a rise in the currency. The 'stock-oriented' approach features the portfolio investor and favours the capital account. Following deviations from expected share values, as they rebalance their portfolios, investors will switch shares so that expected prices re-emerge, this is coloured by currency risks. Using standard Granger-causality tests and impulse response functions, Granger et al. [7] analyses the co-movements of nine countries' exchange rates and their equity market indices during 1987-crash, post-crash and 'Asian flu' eras. The tests are used, in part, to establish a directional flow, but also to consider whether the shocks induced a greater degree of market integration.

Verma and Ozuna [14] find that, when comparing emerging equity markets, risk is related to the currency. Stock markets and currencies in Mexico, Chile and Brazil are negatively related. They point out that equity investments are best viewed as ones of highly uncertain outcomes. Edwards et al. [4] use non-parametric approach to the assessment of Latin American stock market cycles. They find that Latin American market concordance increased post-market liberalization.

Eun and Shim [5] conclude that equity markets of developed economics, including Australia, are information efficient. Their stock markets absorb an innovation from the New York Stock Exchange within two days. Using a common currency, Roca and Buncic [12] do not find long term relations between Australia and South East Asian equity markets following the 'Asian flu', but short term interdependence with Singapore and Hong Kong stock prices is found. Again, the conclusion is efficiency in incorporating information within two days.

Orlov [11] assesses nine Asian and five other currencies in the 1997 'Asian flu' era. Orlov posits that in a tranquil era, exchange rate co-movement may be based on the trend component of the series but there is a shift to the irregular component in the contagious period, suggesting a disproportional increase in the cospectrum during a contagions period at the higher end of the frequency range to reflect greater volatility that fear and speculation precipitate.

Smith [13] uses cross-spectral analysis to assess the impact of the 1987 crash on equity markets of North America and South East Asia. Using coherence among pairs of equity market indices, Smith reveals greater co-movement (intensification) through increased coherence. He uses Wilcoxon's Z to assess whether the two spectra of coherence are drawn from the same population. The Australia and US phase spectrum has generally linear trend with extreme values of zero and π, suggesting a delay of one period (day) of the former behind the latter, which would reflect a timezone difference.

4 The Era of Monetary Confusion

In Copeland [3], there are a series of discussions suggesting that an increase in the nominal money supply will lead to a depreciating exchange rate but the real money stock in the long run remains unchanged. In unusual periods, some presumed stable relations breakdown. With LIBOR becoming sclerotic in mid 2007, banks lost faith in the solvency of other banks and normal lending and, hence, broad money growth in the developed world faded. Perhaps, as an alternative investment class, commodities continued their bull run, peaking in June/July 2008, just before Lehman Bros' collapsed.

In late November 2008, to off-set the pressure from an increase in cash hording and the threat of deflation, the US Federal Reserve started buying $600 billion in mortgage-backed securities and drove interest rates to zero. This should boost the demand for all assets, including equities, both domestic and foreign. Portfolio investors engaged in uncovered interest arbitrage where excess domestic balances are transferred abroad to take advantage of higher returns, the so-called carry trade. Quantitative easing (QE) reduced the value of the US dollar relative to the Real, eventually prompting the currency war claim. Concomitantly, QE may have reinvigorated Brazilian shares immediately with commodities picking up a month later. Minister Mantega imposed a 2 % tax on inflows for equity and fixed-income security purchases in October 2009. QE continued.

5 Methodology

The autocovariance of $X(t)$ in the time domain is the covariance of $X(t)$ against of $X(t-k)$. It is represented as the population [power] spectrum, $s_{XX}(\omega) = \frac{1}{2\pi}\sum_{k=-\infty}^{\infty} \gamma_{XX}(k)e^{-ik\omega}$, in the frequency domain, where $\gamma_{XX}(k)$ is $\text{cov}\{X(t-k), X(t)\}$ and

ω is the periodicy or frequency, measured in radians [8]. The theoretical spectrum divides up a time series into a set of components that are uncorrelated. It reveals the relative power at each frequency corresponding to the variance at each periodicy. A flat spectrum signifies white noise. A general increase in power spectrum values indicates a greater variance of $X(t)$. A specifically elevated spectrum value signifies that that cycle accounts for a relatively a high proportion of the variance of $X(t)$.[1]

The equivalent of covariance and correlation in the time domain are cospectrum and coherence. The population cross spectrum is given by $s_{XY}(\omega) = \frac{1}{2\pi}\sum_{k=-\infty}^{\infty}\gamma_{XY}(k)\,(\cos(\omega k) - i\sin(\omega k))$ where γ_{XY} is the covariance of $\text{cov}\{X(t-k), Y(t)\}$. This can be broken down in the real and imaginary parts, $s_{XY}(\omega) = c_{XY}(\omega) + iq_{XY}(\omega)$, where the cospectrum is defined as $c_{XY}(\omega) = \frac{1}{2\pi}\sum_{k=-\infty}^{\infty}\gamma_{XY}(k)\cos(\omega k)$. If a cospectrum value is large at frequency ω_j, it indicates that cycle accounts for a relatively a high proportion of the covariance of $X(t)$ and $Y(t)$. Squared coherence shows the proportion of a linear relation of $X(t)$ and $Y(t)$ at any frequency [1]. The theoretical squared coherence is given by $C_{XY}^2(\omega) = \frac{|s_{XY}(\omega)|^2}{s_{XX}(\omega)s_{YY}(\omega)}$. An increase in the coherence following a shock is taken to imply at least greater interdependence.

A noise spectrum s_{ZZ} can be estimated as $s_{ZZ}(\omega) = s_{XX}(\omega)\,(1 - C_{XY}(\omega))$ ([10], p. 437). In a sense, it is like unexplained portion of a two variable model.

A phase value is defined as $P_{XY}(\omega) = \tan^{-1}\frac{q_{XY}(\omega)}{c_{XY}(\omega)}$. When phase is a linear function of frequency, the phase diagram can offer an interpretation of 'pure delay' that is independent of ω. The slope in the phase diagram $(-d)$ is the measure of the delay $Y_t = X_{t-d}$. Hilliard, Barksdale and Ahlund [9] seek out linear segments of the phase diagram to reveal leads and lags in beef prices at feeder, live animal, wholesale and retail market levels. Using straight-line segments of the phase diagram, the time delay can be assessed from $dP_{XY}(\omega)/d\omega$. Jenkins and Watts [10] point out that you can get good phase estimates with poor coherence.

As the estimated sample spectra are not consistent, they require a spectral window to be utilised. This entails a moving weighted average across a number of points, which smoothes the spectrum and affects the confidence intervals of the coefficients. As the number of points is increased, the confidence interval expands and the spectrum is smoother.

6 Data

The exchange rate data for the Australian Dollar and Brazilian Real is taken from the European Central Bank. They are based on a regular daily concertation procedure between central banks across Europe and worldwide, which takes place normally at 2.15 p.m. Central European Time (CET). They are valued in US Dollars. The São Paulo Stock Exchange (IBOVESPA) and Australian Stock Exchange (S&P All

[1] So, for example, one would expect variance at periodicies of between 6–10 years would account for a high proportion of the variance in GDP growth.

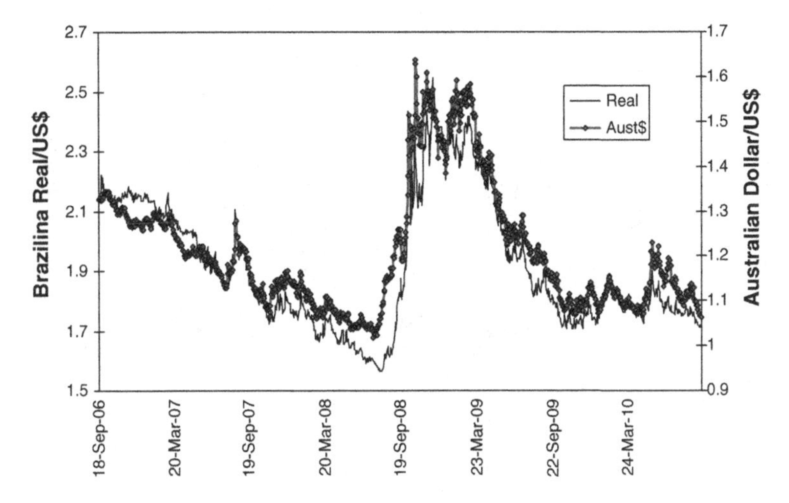

Fig. 2 Currency evolutions

Ordinaries) are taken from the respective stock exchange websites. They are denominated in local currencies and are closing values. There is, therefore, a time delay in measurement associated with timezones. Timezones must be accounted for in any analysis of delays [5]. When it is 4 p.m. in Sydney on Tuesday, it is 2 a.m. in São Paulo; 2 p.m. in Tokyo; 1am in New York; and 5 a.m. in London. Australia and Brazil are so far apart that their markets are not open concurrently. Growth rates are calculated by taking the natural logarithm of the series and then computing the rate of growth as $g_t = \ln p_t - \ln p_{t-1}$. Figures 2–5 illustrate the financial indicators utilised. There appears to be a close link between the Real and the Australian Dollar over the four years of study. Regardless of the impact of the September 2008 crisis, currency movements are similar. Both appreciate against the US Dollar until August 2008; depreciate until around March 2009; appreciate to October 2009; and then stabilise.

As a reference, the Dow Jones Industrial Average is included in Fig. 5 showing the adjusted equity market indices. The two broadly follow the same path as the US stock exchange. The Australian stock exchange rises until November 2007; fall to March 2009; October 2009 and then stabilises. The Brazilian equity index rises until May 2008; decline to October 2008; rise to January 2010 and then stabilises. The turning points for Australia are better attuned that the Brazilian ones to the Dow, consistent with Eun and Shim [5] and Smith [13] finding on efficiency.

Correlations between the indices in growth form are reported in Table 2. Using Spearman's rho, the correlation coefficient concerning the equity markets is 0.231 [0.000], which is lower than for the currencies (0.638 [0.000]). The coefficient for the currency-adjusted correlation is higher than in local currencies (0.46 [0.000]). In other words, there is a closer relationship between equity markets valued in a common currency, which would point to both being influenced by international portfolio investors or international businesses. As the exchange rate depreciated in the post Lehman Bros' period, Brazilian equities decline; for Australia, this change is

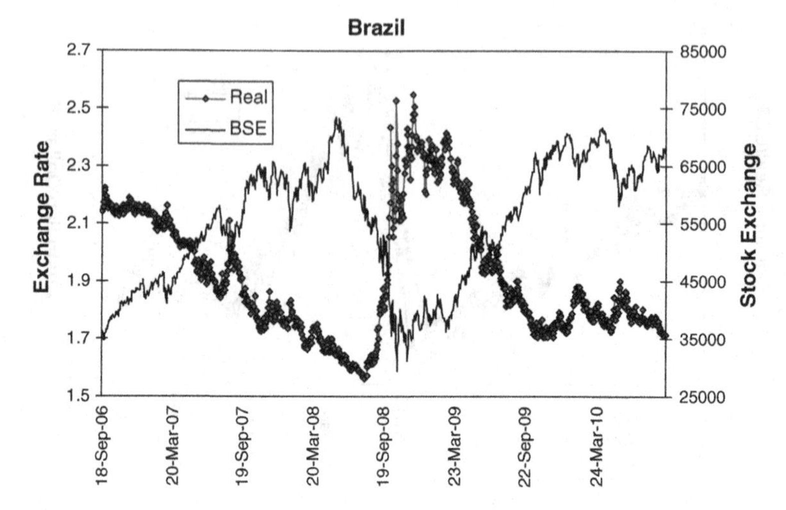

Fig. 3 Currency and equity market evolutions—Brazil

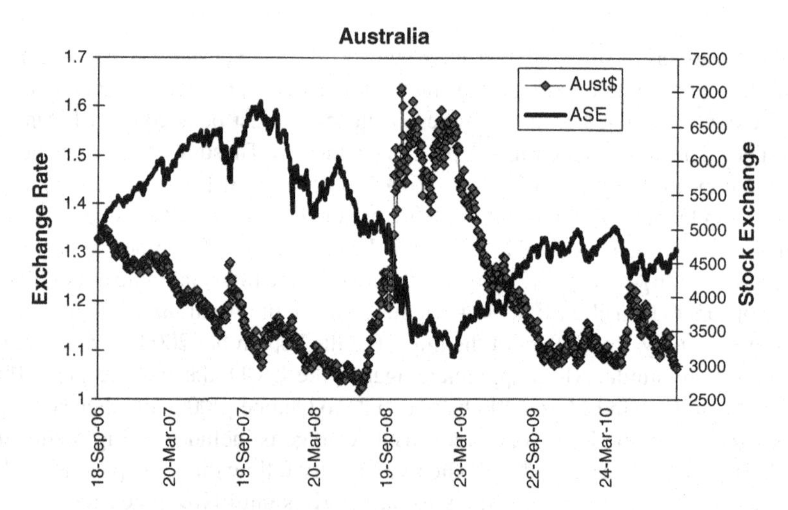

Fig. 4 Currency and equity market evolutions—Australia

not evident. Interestingly, the correlation between the US and Brazil equity market growth (0.566 [0.000]) is above that between 0.46 mentioned above, perhaps reflective of QE funds.

From the measures of standard deviation, skew and kurtosis, displayed in Table 3 the data across the four years, they are least volatile in the benchmark year and in year 3 and most volatile in year 2. Both series in most years exhibit kurtosis. The p-values for the Jarque-Bera test for normality are reported also. The growth data, with one exception, is found to be non-normal.

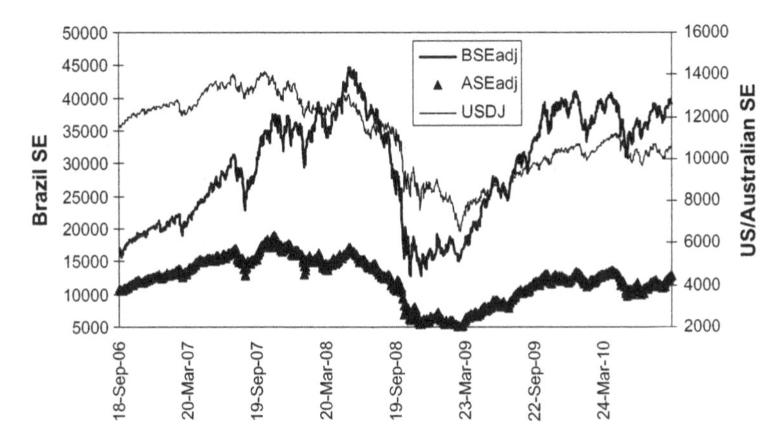

Fig. 5 Equity indices evolutions, common currency

Table 2 Market concordances

Market	Equity	Currency	Equity-currency Brazil	Equity-currency Australia	Equity adjusted
Correlation	0.231**	0.638**	−0.240**	−0.460**	0.460**
p-value	0.000	0.000	0.000	0.000	0.000

∗ indicates significance at the one percent level

Table 3 Descriptive statistics of the equity market indices—common currency

	Year	Mean	Standard deviation	Skewness	Kurtosis	Jarque-Bera
Australian stock exchange	0	0.00120	0.01320	−0.754	7.331	0.000
	1	−0.00091	0.01956	−0.373	1.578	0.000
	2	−0.00018	0.03205	−0.559	2.453	0.000
	3	0.00025	0.01795	−0.218	0.334	0.200
Brazilian stock exchange	0	0.00187	0.01895	−0.621	2.570	0.000
	1	0.00016	0.02375	−0.193	0.768	0.019
	2	0.00022	0.04384	−0.227	3.506	0.000
	3	0.00069	0.01895	−0.164	1.188	0.000
			Std error	0.152	0.303	

As comparisons of spectra are based on particular frequencies, there is a problem with synchronising dates. National holidays and leap years can disrupt the synchronicity. To address this, the data is extended so that there is a value for every weekday, whether it is a working day or not. This is achieved by using the most recent rate in the levels of the data before the non-trading weekday. The four sets of 256 days are taken from 15th September 08 and the equivalent weekday in 2006, 7 and 9.

7 Results

The variables as considered individually across the 4 years and then in pairs. The values for the spectra are estimated by SPSS using a 15-point Parzen lag window for each of the four periods. Marginally greater smoothing did not change inferences. The spectrum comprises 128 (angular) frequencies (*f*), presented as 0.00390625 to 0.5 radians, which corresponds to 256 to 2 periods (days) plus the long run, zero frequency. A radian $\omega_j = 2\,\pi\, f_j$. In addition, there is the long run coefficient at zero frequency.

7.1 Power Spectra and Cospectra

One can adduce in the reference year, there is a high degree of concordance in the lower frequencies. As the cospectrum confirms, this is low past 0.35 radians. Both spectra have three peaks around the same points. In year one, power spectra are greater and of a differing profile to year 0 but still quite similar to each other, as if their profiles have been manipulated by that same forces. Although the similarity between the pairs is not so great, each year's spectra seem different and thse is no tendency obvious confluence of profile over time.

The cospectra point to a general decline in covariance as the frequency increases. Moving from the permanent or trend, to the transitory or irregular component of the spectrum, the two equity markets are less strongly associated. This is not consistent with both being utilised as a substitute or complementary way for speculative purposes.

The coefficients in Table 4 are of the mean of the power spectrum values and the standard deviation. The base year has the lowest values, followed by year 3, then 1 and two. The cospectrum has a differing order, but the base year has the lowest values and year two the highest. There is no general rising trend.

The normalised cumulative power spectra are split into six bands. What emerges from Table 5 is that the weight of the power spectra (volatility) is found in the higher frequencies for both Australia and Brazil. The highest frequencies, with a periodicy of less than $2\frac{1}{2}$ days, account for 14.2 % in the base year in Australian equities, and 18.2 % in Brazilian ones. In addition, a further 15.1 % and 9 % are to be found in the zone $2\frac{1}{2}$ to 3 days, respectively. In this high frequency zone where the equity markets are possibly unrelated, the power spectra accounts for around 29–34 % in the case of Australia and 22–35 % for Brazil.

Contagion should shift distribution of volatility towards the higher frequencies associated with the 'temporary' or 'irregular' component of the adjusted equity market indices (less than $2\frac{1}{2}$ days), particularly in year two. In the peak year of volatility, year 2, the cumulative power spectrum does not indicate that the equity markets have a higher degree of association of volatility in this range than any other. The two equity markets may have been more volatile but the values in Table 5 do not indicate that it is associated specifically with the more speculative end of the spectrum Figs. 6-9.

Table 4 Descriptives of the power spectra and cospectra

	Mean	SD		Mean	SD
AustSE0	0.0266	0.0094	Cospectrum0	0.0187	0.0133
BrazilSE0	0.0550	0.0159			
AustSE1	0.0586	0.0120	Cospectrum1	0.0257	0.0252
BrazilSE1	0.0867	0.0259			
AustSE2	0.1575	0.0243	Cospectrum2	0.1236	0.0764
BrazilSE2	0.2942	0.0860			
AustSE3	0.0495	0.0099	Cospectrum3	0.0287	0.0164
BrazilSE3	0.0550	0.0119			

Table 5 Normalised cumulative power spectra

	Over 1m	2w-1m	1-2w	1w-3d	2.5-3d	<2.5days
	$f < 0.05$	$0.05 > f < 0.1$	$0.1 < f < 0.2$	$0.2 < f < 0.33$	$0.33 < f < 0.4$	$f > 0.4$
AustSE0	0.080	0.155	0.154	0.318	0.151	0.142
AustSE1	0.072	0.100	0.241	0.244	0.136	0.207
AustSE2	0.109	0.116	0.221	0.250	0.111	0.194
AustSE3	0.114	0.131	0.183	0.285	0.113	0.175
BrazilSE0	0.087	0.154	0.191	0.297	0.090	0.182
BrazilSE1	0.080	0.129	0.234	0.203	0.120	0.234
BrazilSE2	0.084	0.143	0.256	0.291	0.088	0.138
BrazilSE3	0.093	0.130	0.240	0.285	0.117	0.135

The average cospectrum value for year 2 in Fig. 10 is higher than for the other years, pointing to greater co-dynamics in the year following Lehman Bros' collapse.

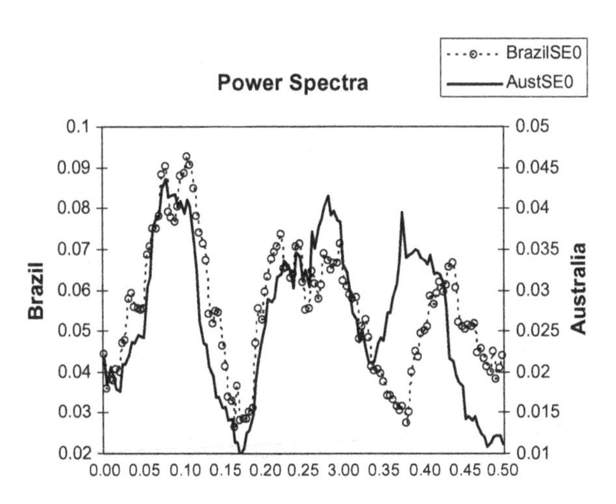

Fig. 6 Power spectra of equity markets, year 0

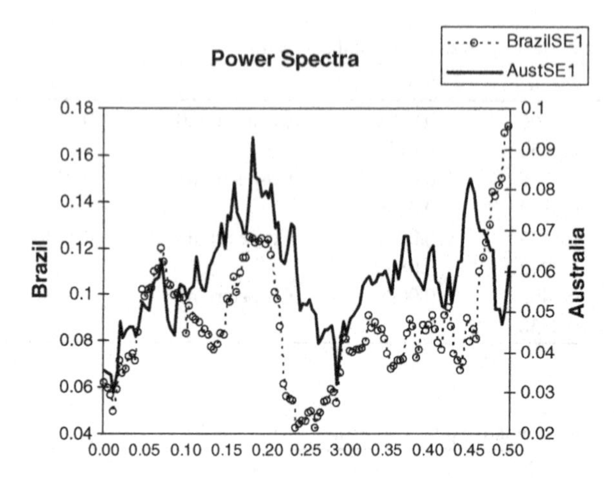

Fig. 7 Power spectra of equity markets, year 1

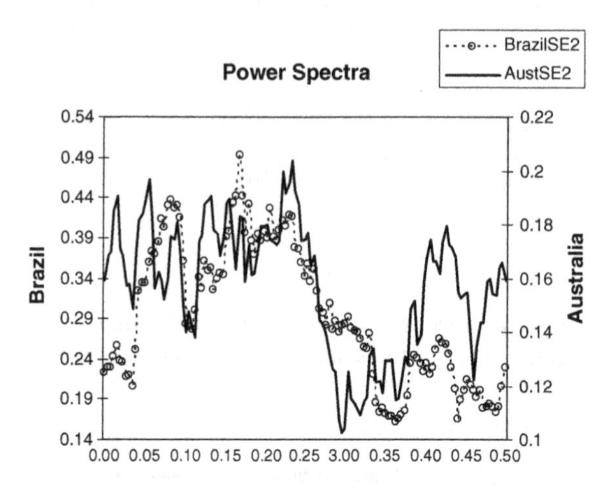

Fig. 8 Power spectra of equity markets, year 2

The general distribution for all years, however, is broadly the same: high positive values in the low frequencies, but beyond 0.33 radians (3 trading days) the values are around zero or negative. Only the average value of the cospectra alters. There is no shift towards the more speculative end of the spectrum.

The error spectrum for all years inclines indicating increasingly distinctive activity between the equity markets at the higher frequencies. They may both suffer from more speculative activity but, as with the cospectra, there is no indication of a change in the co-relationship between the equity markets Figs. 11-15.

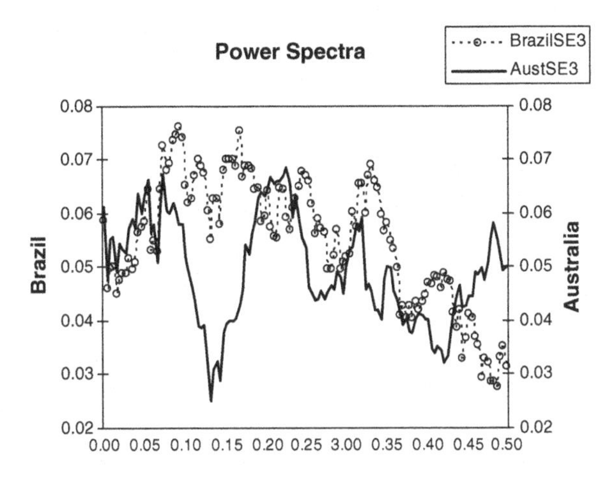

Fig. 9 Power spectra of equity markets, year 3

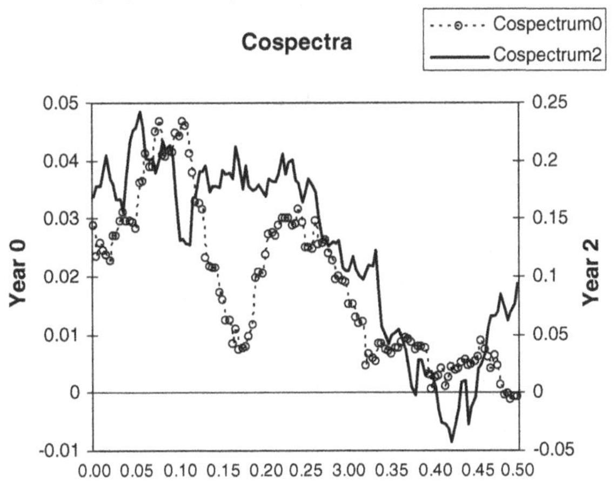

Fig. 10 Cospectra of equity markets, years 0 and 2

7.2 The Coherence Patterns

The coherence spectra are displayed in Fig. 16. The reference year coherence spectrum, Coherence0, declines steadily up to about 0.33 radians, and then the spectrum becomes steeper. This pattern is broadly replicated in year 2. Coherence in years one and three appears to decline more rapidly.

The average value for coherence in the base year is 0.516. In the following three years the corresponding averages are 0.399, 0.519 and 0.46. In spite of the high power and cospectrum values in year 2 which would point to contagion, coherence does not suggest such a structural change. The values in year two are in line with those of year zero rather than one or three Fig. 17.

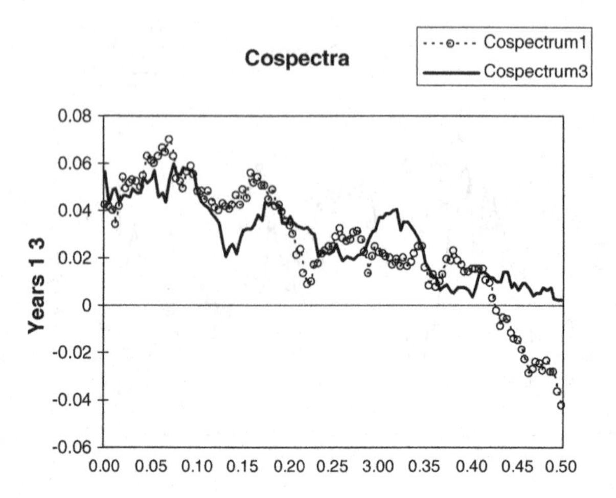

Fig. 11 Cospectra of equity markets, years 1 and 3

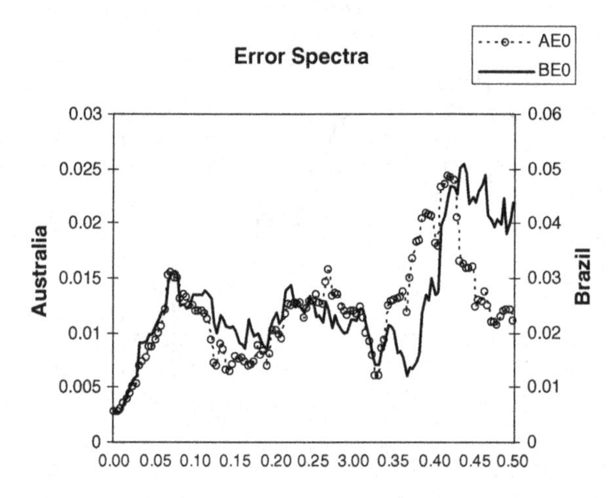

Fig. 12 Error spectra of equity markets, year 0

Smith's test of contagion/intensification entails finding a change in coherence values. As there are four years, there are six possible pairing of coherence. Given the approach taken here, the profile from the base year are contrasted against those from years 1–3, with the expectation that they are not different. In Table 6, the *p*-values from the Wilcoxon and Sign tests point to year 2 having the profile similar to that of the base year ($p = 0.841$; 0.291), with a possibility of years 2 and 3 being indistinct.

A Friedman Test value of 37.214 [0.000] points to a rejection of the hypothesis that the coherence values from all four years are drawn from the same pool. However, the corresponding values excluding year 1 indicates that they are drawn from the

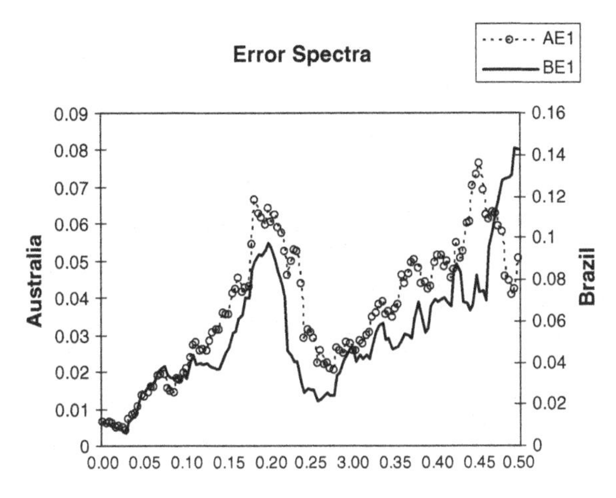

Fig. 13 Error spectra of equity markets, year 1

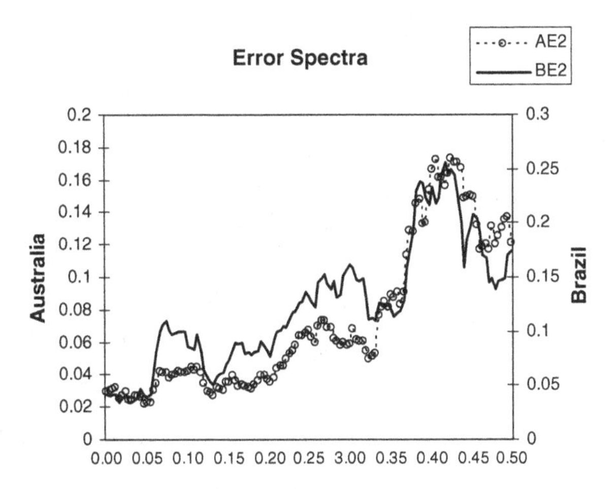

Fig. 14 Error spectra of equity markets, year 2

same pool $\{\chi^2 = 4.388[0.111]\}$. This suggests the year before Lehman Bros' was the exception to a stable rule.

7.3 Phase

The last consideration is that of phase. Yet again, there is stability. Start and finish points of approximately linear portions of the phase spectra are reported in Table 7. Accompanying them are the gradients that can be interpreted directly in hours as

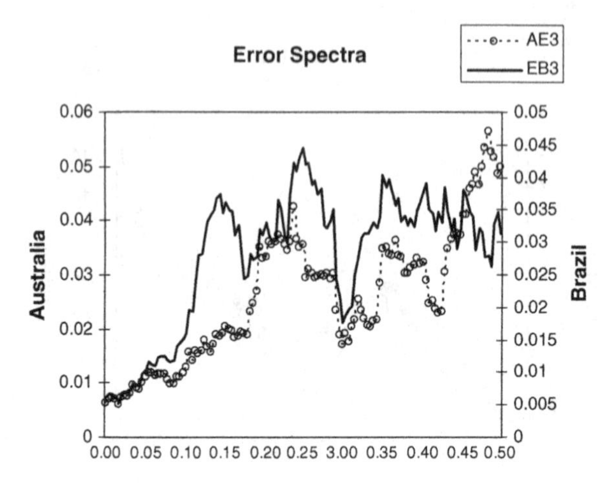

Fig. 15 Error spectra of equity markets, year 3

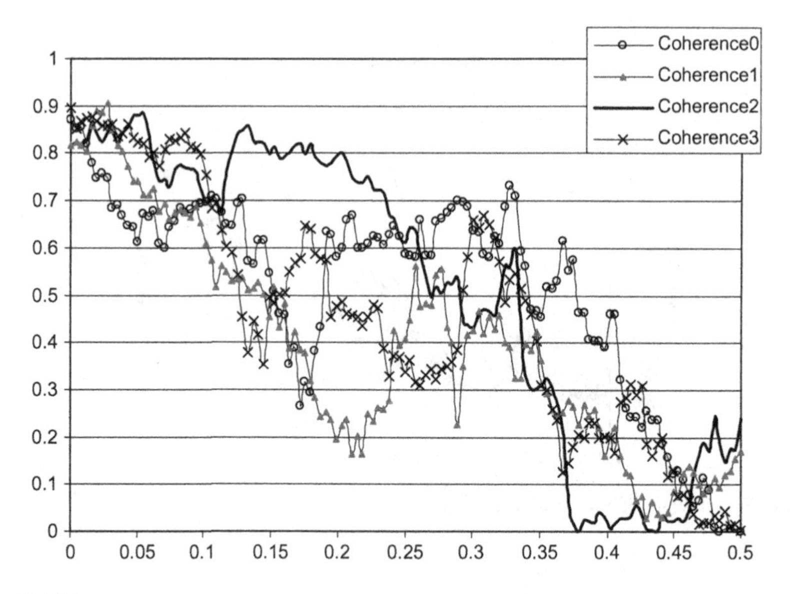

Fig. 16 Coherence

leads and lags in the time domain. The phase angle for all four years appears to be linear function of frequency up to when the cospectrum approaches zero. In the base year, the angle points to a delay of 14 h, which is the timezone differential between São Paulo and Sydney. The corresponding delays are smaller for the later years, suggesting the Australian market is more information efficient. However, taking into account the confidence interval for phase, with the exception of year 2, the post Lehman's collapse year, the delays across the lower frequencies, with periodicies up

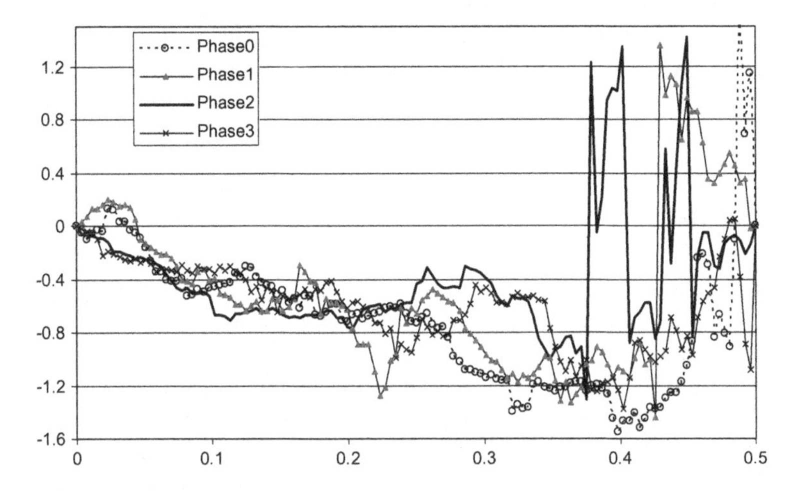

Fig. 17 Phase diagrams

Table 6 Smith tests of coherence values

Year	0		1		2	
	Wilcoxon	Sign	Wilcoxon	Sign	Wilcoxon	Sign
1	0.000	0.000				
2	0.841	0.291	0.000	0.000		
3	0.001	0.035	0.000	0.003	0.002	0.291

Table 7 Leads and lags

	f start	*f* finish	Delay (h)	95 % confidence interval (h)
Trend portion				
Phase0	0	0.4023	−14	±3
Phase1	0	0.4258	−12	±5
Phase2	0	0.3477	−10	±3
Phase3	0.0195	0.3984	−10	±4
Irregular portion				
Phase20	0.4023	0.5	9	n/a
Phase21	0.4336	0.5	−10	n/a
Phase22	0.3477	0.5	0	n/a
Phase23	0.3984	0.5	42	n/a

to 3 days, are consistent with the timezone differences. This implies that there is no difference in the absorption of information: the markets are relatively information efficient.

Higher frequencies provide distinct and, perhaps, unreliable stories. This could be related to the very low level of coherence and the irregular nature of currency or equity market speculation. Figures are reported.

8 Conclusion

This work explores Brazilian-Australian equity market concordance using spectral techniques. The aim is to reveal evidence for a shift in the relationship reflecting either a trend or a shock.

It appears that the Lehman Bros' crisis masks a commodity price collapse. Turbulence is evident in the power and cospectra, but not in phase or coherence. Indeed, the tests of difference highlights one unusual year, 2007/8. Furthermore, there appears to be no evidence in any spectrum other than in the power spectrum of year 1 that there is a shift towards the higher frequencies, consistent with short-term speculative investment. Thus, rather than being subject to contagion, the concordance between Brazil's and Australia's markets in year 3 is as it was in the reference year. Indeed, there is no difference in the absorption of information. Any delay is related to time-zone differences: the markets are relatively information efficient.

The movement of cheap money from loose monetary policy zones and capital restrictions have not altered the degree of equity co-movement, nor is there evidence of portfolio or speculative trading affecting the profile of volatility. When money market dried up Northern Rock collapsed and commodity prices fell, the distinctive period is the year before the crisis. Overall, it appears that portfolio investment does not trump trade with commodity-driven equity markets. In this era, currencies were appreciating. The currency valuation is not the key determinate of stock market performance. More likely, it was the price of commodities that drove equities.

References

1. Bartels, C.: Economic Aspects of Regional Welfare, Income Distribution and Unemployment. Martinus Nijhoff Social Sciences Division, Leiden (1977)
2. Bartram, S., Bodnar, G.: No place to hide: the global crisis in equity markets in 2008/2009. J. Int. Money Finan. **28**(8), 1246–1292 (2009)
3. Copeland, L.: Exchange Rates and International Finance, 4th edn. Prentice Hall, Harlow (2005)
4. Edwards, S., Biscarri, J.: Stock market cycles, financial liberalization and volatility. J. , Perez de Gracia Int. Money Finan. **22**(7), 925–955 (2003)
5. Eun, C., Shim, S.: International transmission of stock market movements. J. Finan. Quant. Anal. **24**(2), 241–256 (1989)
6. Forbes, K., Rigobon, R.: No contagion, only interdependence: measuring stock market comovements. J. Finan. **57**(5), 2223–2261 (2002)
7. Granger, C., Huang, B.-N., Yang, C.-W.: A bivariate causality between stock prices and exchange rates: evidence from recent Asian flu. Q. Rev. Econ. Finan. **40**(3), 337–354 (2000)
8. Hamilton, J.: Time Series Analysis. Princeton University Press, Princeton (1994)

9. Hilliard, J., Barksdale, H., Ahlund, A.: A cross-spectral analysis of beef prices. Am. J. Agric. Econ. **57**(3), 309–315 (1975)
10. Jenkins, G., Watts, D.: Spectral Analysis and Its Applications. Holden-Day, London (1968)
11. Orlov, A.: A cospectral analysis of exchange rate co-movements during the Asian financial crisis. J. Int. Finan. Markets Inst. Money **19**(5), 742–758 (2009)
12. Roca, E., Buncic, D.: Equity market price interdependence between Australia and the Asian tigers. Int. J. Bus. Stud. **10**(2), 61–74 (2002)
13. Smith, K.: Pre- and post-1987 crash frequency domain analysis among Pacific rim equity markets. J. Multinatl. Finan. Manage. **11**(1), 69–87 (2001)
14. Verma, R., Ozuna, T.: Are emerging equity markets responsive to cross-country macroeconomic movements?: evidence from Latin America. J. Int. Finan. Markets Inst. Money **15**(1), 73–87 (2005)
15. Wolf, M.: Currencies clash in new age of beggar-my-neighbour. The Financial Times. http://www.ft.com/cms/s/0/9fa5bd4a-cb2e-11df-95c0-00144feab49a.html (2010). 29 Sep 2010, Accessed 29 March 2012

[illegible] A cross-spectral analysis [illegible]
[illegible] (1959).
16. [illegible] Spectral Analysis and Its Applications. Holden-Day, [illegible]
[illegible] rate [illegible]
[illegible] 74 [illegible]
[illegible]
[illegible]
[illegible]
[illegible]
[illegible]
[illegible]
[illegible]
[illegible]
[illegible]

Chapter 4
Probabilistic European Country Risk Score Forecasting Using a Diffusion Model

R. Cervelló-Royo, J.-C. Cortés, A. Sánchez-Sánchez, F.-J. Santonja, R. Shoucri and R.-J. Villanueva

Abstract Over the last few years, global crisis has shaken confidence in most European economies. As a consequence, a lack of confidence has spread amongst European countries leading to Europe's financial instability. Therefore, forecasting the next future of economic situation involves high levels of uncertainty. In this respect, it would be interesting to use tools which allow to predict the trends and evolution of each country's confidence rating. The Country Risk Score (CRS) represents a good indicator to measure the current situation of a country regarding measures of economic, political and financial Risk in order to determine country Risk ratings. CRS is underscored by Euromoney Agency and is calculated by assigning weights to these measures. In this contribution, we present a diffusion model to study the dynamics of the CRS in 27 European countries which considers both the endogenous effect of each country policies and the contagion effect among them. The model

---

R. Cervelló-Royo (✉)
Department of Economics and Social Sciences, Universitat Politècnica de València, Valencia, Spain
e-mail: rocerro@esp.upv.es

J.-C. Cortés · A. Sánchez-Sánchez · R.-J. Villanueva
Instituto Universitario de Matemática Multidisciplinar, Universitat Politècnica de València, Valencia, Spain
e-mail: jccortes@imm.upv.es

A. Sánchez-Sánchez
e-mail: alsncsnc@posgrado.upv.es

R.-J. Villanueva
e-mail: rjvillan@imm.upv.es

F.-J. Santonja
Departamento de Estadistica e Investigación Operativa, Universitat de València, Valencia, Spain
e-mail: santonja@uv.es

R. Shoucri
Royal Military College, Kingston, ON, Canada
e-mail: shoucri-r@rmc.ca

V. K. Mago and V. Dabbaghian (eds.), *Computational Models of Complex Systems*, Intelligent Systems Reference Library 53, DOI: 10.1007/978-3-319-01285-8_4,
© Springer International Publishing Switzerland 2014

depicts quite well the evolution of the CRS despite jumps and uncertainty in the data within some periods. Furthermore, it should be noted a downward trend in the CRS dynamics for almost all European countries in the next year.

1 Introduction and Motivation

The international global crisis has shaken confidence not only in the euro but also in European countries. Over the last four decades, the European countries have experienced a strong convergence, taking in poor countries and helping them become high-income economies. However, since 2008 Northern countries, Continental and some Baltic Economies have done well, whereas Southern Europe Economies and other economies have not. As a consequence, an increasing loss of confidence in these countries has had a strong negative contagion effect on the rest leading all of them to an uncertain scenario from an economical point of view.

This loss of confidence has a strong influence in the capital flows [10, 22] and the investors' attitude towards some countries [5, 7, 8, 11, 15, 18]. Thus, it can be stated that financial crises affect capital flows and investors' attitudes in a reversal way when a country shifts abruptly from a good equilibrium with a low country-specific risk premium to a bad equilibrium with a high country-specific risk premium and, as a consequence, no foreign credit and investment in sovereign debt.

Core eurozone sovereigns and confidence have deteriorated sharply since the global financial crisis, with Germany, Austria and The Netherlands among the countries to register declines since 2008 drawn by other European countries whose situation has put in doubt the current European cohesion and their capacity to repay their debts. Traditional concepts of risk, solvency and liquidity allow us to understand this topic. Therefore, being able to assess the creditworthiness of countries is a key issue.

Country risk has become a topic of major concern for the international financial community over the last two decades. The importance of country ratings is underscored by the existence of several major country risk rating agencies [14]. For this purpose, Country Risk Scores (CRS) are built in order to measure several factors, both quantitative and qualitative. In our study we use the CRS underscored by the Euromoney Agency [4], which combines the following categories: political risk, economic performance, debt indicators, structural assessments, access to bank finance/capital markets and credit ratings. The categories involve the following information:

- C1: Political risk: corruption; government non-payments/non-repatriation; government stability; information access/transparency; institutional risk; regulatory and policy environment.
- C2: Economic performance: bank stability/risk; economic GNP (Gross National Product); employment/unemployment; monetary policy/currency stability; government finances.

- C3: Debt indicators: total debt stocks to GNP, debt service to exports and current account balance to GNP.
- C4: Structural assessments: demographics; hard infrastructure; labour market/ industrial relations; soft infrastructure.
- C5: Access to bank finance/capital markets: country's accessibility to international markets.
- C6: Credit ratings: nominal values are assigned to sovereign ratings from Moody's, Standard & Poor's and Fitch IBCA.

Thus, CRS can represent a good indicator of the current situation of a country regarding measures of economic, political and financial risk in order to determine country risk ratings. In the case of Euromoney Agency, the overall ECR (Euromoney Risk Score) is obtained by assigning to the six categories the following weights:

- 3 qualitative expert opinions: political risk (30 % weighting), economic performance (30 % weighting) and structural assessment (10 % weighting).
- 3 quantitative values: debt indicators (10 % weighting), credit ratings (10 % weighting), access to financial markets (10 % weighting).

When talking about financial crises there is a lot of literature that takes into account several reasons for crises to appear in clusters [5, 6, 11, 18]. Therefore, we consider that in Europe a crisis in one European country may focus investors' attention on other European countries with similar trends and general structural similarities and vulnerabilities. This effect is widely known as *common weakness contagion* [1, 2, 9, 21]. That is the main reason an analysis cluster has been performed in our study. The contagion is usually modelled using epidemiological and/or diffusion techniques. Both are in close connection and let us study the dynamics of CRS using these mathematical techniques. Our objective is to predict the CRS trends over the next year, providing prediction tools for policy makers. Thus, with these tools, policy makers are able to design strategies, simulate different situations and analyse the effect of changes in order to improve the economic situation.

This chapter is organized as follows: Sect. 2 is addressed to introduce the available data, to perform an analysis cluster and to construct and justify the mathematical model used to describe the dynamics of the CRS of 27 European countries. First part of Sect. 3 is devoted to model parameter estimation and CRS deterministic punctual forecasting over the next few months. Since uncertainty and variability are the rules when dealing with modelling real problems, in the second subsection, we complete our predictions by means of confidence intervals. Conclusions are drawn in Sect. 4.

2 Modelling

This section is addressed to construct and justify the mathematical model used to describe the dynamics of the CRS of the 27 European countries which, according to previous exposition, provides a reliable economic indicator of the current situation

Table 1 Clustering analysis obtained by the non-hierarchical clustering technique

i	Country name	Cluster	*i*	Country name	Cluster
1	Luxembourg	1	8	United Kingdom	2
2	Denmark	1	9	France	2
3	Sweden	1	10	Cyprus	2
4	Finland	1	11	Belgium	2
5	Netherlands	1	12	Malta	2
6	Germany	1	13	Czech Republic	2
7	Austria	1	14	Slovenia	2
			15	Slovak Republic	2
			16	Italy	2
			17	Poland	2
			18	Estonia	2
19	Spain	3	24	Hungary	4
20	Ireland	3	25	Bulgaria	4
21	Portugal	3	26	Latvia	4
22	Lithuania	3	27	Romania	4
23	Greece	3			

Table 2 Initial CRS data corresponding to March 31, 2011 [4]

Country	CRS initial value	Country	CRS initial value
Luxembourg	88.27	United Kingdom	82.14
Denmark	88.80	France	82.02
Sweden	88.93	Cyprus	79.95
Finland	88.55	Belgium	80.22
The Netherlands	88.20	Malta	75.67
Germany	84.52	Czech Republic	74.73
Austria	85.80	Slovenia	78.71
		Slovak Republic	76.27
		Italy	74.01
		Poland	70.47
		Estonia	61.63
Spain	72.27	Hungary	61.32
Ireland	77.87	Bulgaria	59.93
Portugal	73.81	Latvia	55.71
Lithuania	60.39	Romania	53.52
Greece	60.33		

of every European country (Table 1). To perform the study we have considered a total of 21 available values of CRS's for every country corresponding to different dates starting from March 31, 2011 to March 19, 2012 [4]. Table 2 collects these numerical values corresponding to the starting date.

2.1 Analysis Cluster

As we said previously, in finance, and especially in country risk assessment, it is useful to group the different countries sharing similar economic characteristics. Therefore, before constructing our diffusion mathematical model, we have performed a clustering analysis. The clustering technique allows us to gather the different countries into homogeneous groups. In order to deal with this task, we have used the Non-Hierarchical clustering (also called k-means clustering) [13, 16]. This method separates n observations into k clusters in which each observation belongs to the cluster with the nearest mean. We have grouped all the European countries into four clusters considering the available data corresponding to the six categories C1–C6 introduced in the previous section. The results are reported in Table 1.

The first cluster gathers the Nordic and Continental safest European economies, it should be also remarked that Denmark and Sweden do not belong to the Eurozone. The Second cluster gathers Continental European Countries and United Kingdom (UK); except for UK, all of them have the same currency (Euro) or are in an adaptation process; it should be also remarked the presence of Estonia, as the Baltic country which has done better since it joined the Euro in 2004. The third cluster includes most of the European Economies which have been going through sovereign debt crises or have been bailed out by the European Union; all of them belong to the Eurozone area, except for Lithuania which is in an adaptation process. The fourth cluster includes the weakest European Economies; none of them has adopted the Euro yet.

In the following, for our dynamic model, we are going to assume that the obtained clustering does not change over the time. This hypothesis is reasonable because we are going to predict CRS evolution in a short time and there will be very few countries that may move out from one cluster to another in the period studied. Furthermore, the obtained clustering represents quite accurately the current economic European zones.

2.2 Mathematical Model

Once the clusters have been established, we propose a type-diffusion dynamic mathematical model to study the evolution of the CRS of each European country. Type-diffusion dynamic models have demonstrated to be powerful tools to study a wide range of applied problems in different areas including Economics and its related fields [12, 17, 23]. Diffusion models also apply successfully in Epidemiology to describe the spread of diseases [3]. Although very complex models have been proposed based on this approach, all of them are mainly based on the following pattern:

$$x'(t) = \beta(t)x(t), \tag{1}$$

where $x(t)$ represents the magnitude under modelling (in our case, the CRS to each European country at time t), $x'(t)$ denotes the derivative with respect to the time t (it may also be denoted by $\frac{dx(t)}{dt}$) and $\beta(t)$ is the so-called time-dependent diffusion coefficient, which may involve the unknown $x(t)$. This coefficient is basically what differentiates one specific model from another. As we shall see in detail later, our model considers $\beta(t)$ as a linear function which can be decomposed into two factors. The first one represents, through CRS, the autonomous economic behavior of each country and, the second one, the contagion effect for loss of confidence both between and within clusters for each country.

For the sake of clarity in the model setting, we will identify each one of the 27 European countries with the index $i = 1, 2, \ldots, 27$ according to the same order obtained after clustering (see Table 1). Now, it is convenient to denote j_k and J_k, $1 \leq k \leq 4$ as follows:

$$\begin{array}{ll} 1 \leq j_1 \leq J_1 = 7, & 1 \leq j_2 \leq J_2 = 11, \\ 1 \leq j_3 \leq J_3 = 5, & 1 \leq j_4 \leq J_4 = 4, \end{array} \qquad \sum_{k=1}^{4} J_k = 27, \tag{2}$$

where J_k is the number of countries in cluster k. As we have already pointed out, we assume that CRS variation rate of a country, modelled by its derivative $C_i'(t)$, $1 \leq i \leq 27$, is a mixture of an autonomous term, related to its endogenous economic politics, and several diffusion terms, related to the exogenous economic influences of other European countries belonging to same or different cluster. In the following, we describe in detail these terms:

- **Autonomous behavior**: Each European country develops endogenous politics that may result in an increase or decrease of its own CRS. We model this autonomous behavior by $\alpha_i C_i(t)$. The coefficient α_i can take positive or negative values which are identified, respectively, with suitable and unsuitable domestic politics.
- **Transmission behavior**: In an economically globalized world, the economic situation of a specific country could eventually influence on other countries. This statement makes even more realistic for a set of countries sharing the same economic regulations such as European countries. Initially, this might motivate to establish a type-diffusion model by considering a full contagion between each pair of countries. However this approach entails the introduction of a large number of parameters what would make the model parameter estimation computationally unfeasible. In order to reduce the number of model parameters, we take advantage of previous clustering classification to consider as a balanced CRS' indicator of each cluster the following average value:

$$\overline{C}_k(t) = \frac{1}{J_k} \sum_{j_k=1}^{J_k} C_{j_k}(t), \quad 1 \leq k \leq 4. \tag{3}$$

Based on the CRS, we propose to model the influence of the economic policies of the countries belonging to the cluster k, $1 \leq k \leq 4$, on the i-th European country, $1 \leq i \leq 27$, according to the following term (t measured in years):

$$\beta_{k(i),k} C_i(t) \left(\overline{C}_k(t) - C_i(t)\right), \quad \beta_{k(i),k} > 0, \ 1 \leq k \leq 4, \tag{4}$$

where the first subindex $k(i)$ of coefficient $\beta_{k(i),k}$ denotes the cluster to which the i-th country belongs:

$$k(i) = \begin{cases} 1 \text{ if } 1 \leq i \leq 7 = J_1, \\ 2 \text{ if } 8 \leq i \leq 18 = J_1 + J_2, \\ 3 \text{ if } 19 \leq i \leq 23 = J_1 + J_2 + J_3, \\ 4 \text{ if } 24 \leq i \leq 27 = J_1 + J_2 + J_3 + J_4. \end{cases} \tag{5}$$

In accordance with the general pattern for diffusion models (1), in our case the diffusion coefficient in (4) is given by $\beta(t) = \beta_{k(i),k} \left(\overline{C}_k(t) - C_i(t)\right)$ which depends on the unknowns $C_i(t)$. The two factors of this coefficient can be interpreted as follows:

- The term $\overline{C}_k(t) - C_i(t)$ measures the CRS difference (positive or negative) between the i-th country at time t, denoted by $C_i(t)$, and the CRS average of countries belonging to cluster k at the same time t, denoted by $\overline{C}_k(t)$. So, we assume that this factor contributes positively (negatively) to the transmission term (4) when the i-th country has a CRS lower than the CRS average of countries in cluster k. Thus, we suppose that countries with lower (higher) CRS than those belonging to a cluster with higher (lower) average CRS tend to increase (decrease) their CRS influenced by the countries belonging to cluster k.
- The factors responsible of the contagion effect are embedded in the $\beta_{k(i),k}$ coefficients, and they modulate the weight of the differences $\overline{C}_k(t) - C_i(t)$ in the transmission terms. Notice that, once the subindex k is fixed, we are assuming that this coefficient $\beta_{k(i),k}$ is the same for every country belonging to same cluster $k(i)$. This reduce the total number of $\beta_{k(i),k}$ up to 16.

In order to consider the possible influence of countries belonging to every cluster, we generalize the transmission term (4) as follows:

$$\sum_{\substack{1 \leq i \leq 27 \\ 1 \leq k(i) \leq 4}} \beta_{k(i),k} C_i(t) \left(\overline{C}_k(t) - C_i(t)\right), \quad \beta_{k(i),k} > 0, \ 1 \leq k \leq 4. \tag{6}$$

Taking into account the previous exposition that includes both autonomous and transmission behavior, we propose the following type-diffusion continuous dynamic model, based on a coupled system of 27 nonlinear differential equations, to study the dynamic evolution of the CRS's European countries:

$$C_i'(t)=\alpha_i C_i(t)+\sum_{\substack{1\leq i\leq 27\\1\leq k(i)\leq 4}}\beta_{k(i),k}C_i(t)\left(\overline{C}_k(t)-C_i(t)\right),\ 1\leq i\leq 27,\ 1\leq k\leq 4, \tag{7}$$

3 Prediction Over the Next Few Months

This section is divided into two parts. The first one, is devoted to model parameters estimation and CRS deterministic punctual forecasting, monthly, over the next 12 months. Since uncertainty and variability are the rules when dealing with modelling real problems, in the second subsection, we complete our predictions by means of confidence intervals.

3.1 Parameters Estimation and Deterministic Punctual Predictions

As we have previously pointed out, this subsection is firstly addressed to estimate the parameters of model (7). This task has been performed by fitting the model in the mean square sense to the available data. Computations have been carried out with *Mathematica 8.0* [19].

The system of differential equations (7) is numerically solved by taking as initial conditions the CRS data of March 31, 2011 (corresponding to $t=0$), i.e., according to Table 2, $C_1(0)=88.27$, $C_2(0)=88.8,\ldots$ and $C_{27}(0)=53.52$. Tables 3 and 4 report the estimation of the autonomous (α_i) and contagion ($\beta_{k(i),k}$) model parameters. For the sake of clarity, in Fig. 1 we have depicted the estimation of the autonomous model parameters in four plots according to previous clustering. This allows us to observe a common pattern of these estimations depending on clustering which strengthens our approach.

Also, looking at Table 4, we notice that the estimated transmission parameters suggest that countries in cluster $k=1$ have a great influence on themselves whereas the cluster $k=3$ influences on the rest. This last statement corresponds with the concern in the European Union lately with the economic situation of countries in cluster $k=3$ (Spain, Ireland, Portugal, Greece and Lithuania).

From the parameter estimations shown in Tables 3 and 4, we have obtained punctual predictions of CRS for every European country. The solid lines plotted in Fig. 2 represent these predictions for several selected countries of each cluster as representatives of each cluster, namely, cluster 1: Germany and Finland; cluster 2: France and Estonia; cluster 3: Spain and Greece; cluster 4: Hungary and Romania. Notice that the available data of CRS are represented by points in Fig. 2.

Table 3 Estimation of the autonomous model parameters α_i, separated by clusters

Country	α_i	Country	α_i
Luxembourg	1.192	United Kingdom	0.277
Denmark	0.908	France	0.298
Sweden	0.875	Cyprus	0.078
Finland	0.903	Belgium	0.227
The Netherlands	0.817	Malta	0.215
Germany	0.548	Czech Republic	0.218
Austria	0.457	Slovenia	0.226
		Slovak Republic	0.210
		Italy	0.064
		Poland	0.174
		Estonia	0.198
Spain	−0.041	Hungary	−0.017
Ireland	−0.026	Bulgaria	−0.225
Portugal	−0.148	Latvia	−0.343
Lithuania	0.002	Romania	−0.409
Greece	−0.500		

Table 4 Estimated values of the contagion model parameters

$\beta_{k(i),k}$	$k = 1$	$k = 2$	$k = 3$	$k = 4$
$k(i) = 1, 1 \le i \le 7$	0.099	0	0.028	3.2×10^{-5}
$k(i) = 2, 8 \le i \le 18$	6.08×10^{-6}	9.79×10^{-6}	0.015	0
$k(i) = 3, 19 \le i \le 23$	6.77×10^{-6}	0	4.0×10^{-3}	1.4×10^{-6}
$k(i) = 4, 24 \le i \le 27$	2.7×10^{-4}	0	0.066	5.6×10^{-5}

The value of model parameter $\beta_{k(i),k}$ measures the contagion effect transmitted by the countries belonging to cluster k ($1 \le k \le 4$) on country i ($1 \le i \le 27$) belonging to cluster $k(i)$ according to assignment (5). Figures indicate that countries in cluster $k = 1$ have a great influence on themselves (0.099) whereas the countries in cluster $k = 3$ influence on the rest (column $k = 3$)

3.2 *Introducing Uncertainty into the Model: Predictions Over the Next Few Months Using Confidence Intervals*

Randomness can be attributed not only to sampling errors in the data but also to the inherent complexity of the phenomenon under study. This statement particularly holds in dealing with economic problems. Therefore, in order to complete the punctual prediction of CRS provided previously, it is more realistic to construct predictions by confidence intervals. To calculate these intervals, let us use an adaptation of the statistical technique usually referred to as cross-validation or rotation estimation. Cross-validation is a versatile statistical method of evaluating and comparing learning algorithms by dividing data into two segments: one used to learn or train the model and the other used to validate the model. Apart from its basic formulation,

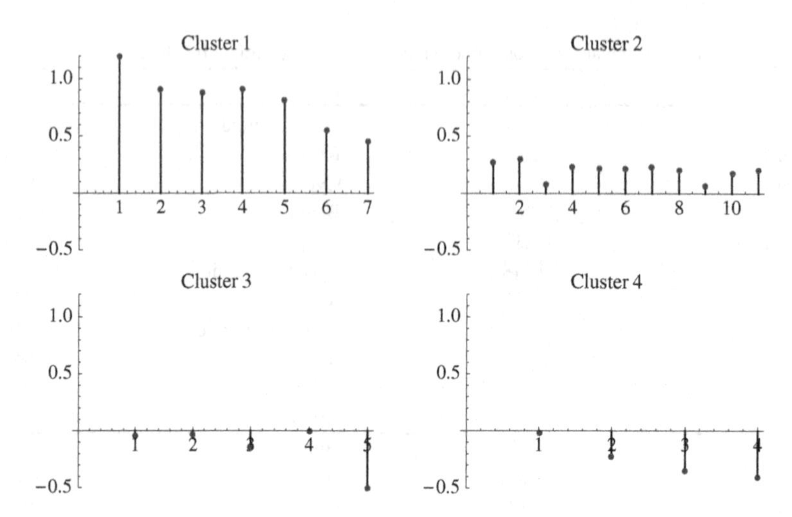

Fig. 1 Representation of the estimation of the autonomous model parameters per cluster. Notice the common pattern of these estimations depending on clustering. The countries are in the same order as they appear in Table 3

several variations of cross-validation have been proposed to get other statistical goals including the estimation of model parameters.

The version of the cross-validation process we propose is the following: we have 27 CRS data, one for each country, for 21 different time instants between March 31st, 2011 and March 19th, 2012; we take the CRS data corresponding to K time instants among the 21 available, number K to be determined, where data of March 31, 2011 (initial condition of the model) are always included; fit the model with the K data taken, obtaining an estimation of model parameters; substitute the parameters estimated into the model and solve the model numerically; compute model outputs in the known 21 time instants and monthly, from April 1st, 2012 until March 1st, 2013, these last ones for predictions. If we repeat this process around $5,000$ times, we will be able to build 90 % confidence intervals by computing percentiles 5 and 95 of model outputs in each one of the aforementioned time instants in order to provide probabilistic predictions.

Now, we should find an appropriate K such that most of the 27 CRS data points lie into the 90 % confidence intervals. After some computational tests:

- **Step 1**: We find out that taking subsets of $K = 5$ (one of them is always the initial condition), we will have 4,845 different subsets of $K = 5$ CRS data elements.
- **Step 2**: For each one of the 4,845 subsets, we fit the model (7) with this subset using the Nelder-Mead algorithm [20] taking an initial polyhedra around the model parameters estimated in Sect. 3.1 (the ones that best fit the model). Thus, we get $4,845$ sets of model estimated parameters that best fit subsets of $K = 5$ CRS data elements.

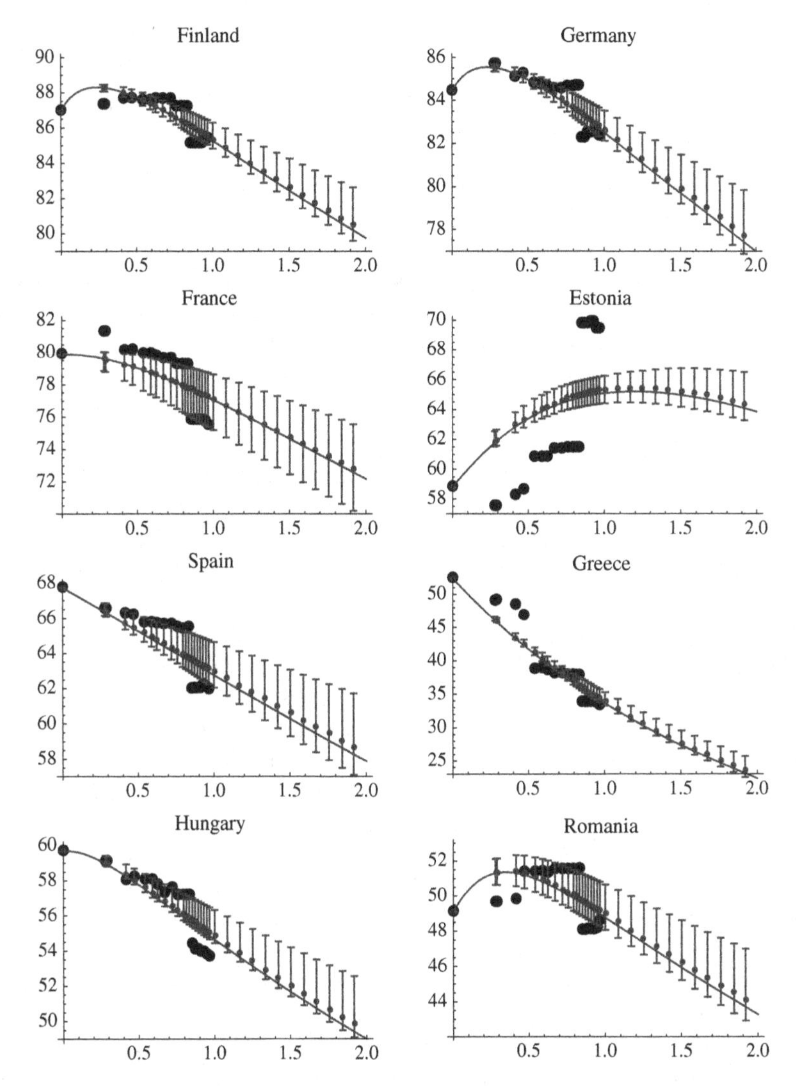

Fig. 2 Probabilistic country risk score forecasting for selected European countries belonging to each cluster (Cluster 1: Finland and Germany. Cluster 2: France and Estonia. Cluster 3: Spain and Greece. Cluster 4: Hungary and Romania). The *solid line* is the model solution for the parameters that best fit the model (deterministic model). The *points* are the known CRS data from March 31st, 2011 to March 19th, 2012. Except for Estonia, there are decreasing trends in the CRS for the next year. CRS values appear in the *vertical axis*, and the time (in years), starting on March 31, 2011, in the *horizontal axis*

- **Step 3**: We substitute each one of the 4,845 sets of model estimated parameters into the model and solve it numerically.

- **Step 4**: Compute the output of each one of the 4,845 model solutions in the 21 time instants and monthly, from April 1st, 2012 until March 1st, 2013.
- **Step 5**: For each time instant we take the 4,845 model outputs and compute the mean and the 90 % confidence intervals by percentiles 5 and 95.

The obtained results of the above process can be seen graphically in Fig. 2 for Finland and Germany in cluster 1, France and Estonia in cluster 2, Spain and Greece in cluster 3, and Hungary and Romania in cluster 4, as representatives of each cluster. In Fig. 2 we show the evolution of the CRS of the mentioned countries, the model deterministic prediction (solid line) and the probabilistic prediction (vertical segments) corresponding to 90 % confidence intervals in the 21 time instants between March 31st, 2011 and March 19th, 2012 and monthly, from April 1st, 2012 until March 1st, 2013. The points are the known CRS data.

According to the obtained results, we should remark the difficult task of forecasting the complex economic dynamics of 27 European countries considering the current international scenario, added to the uncertainty in the Eurozone area. The CRS data behavior reflect these circumstances, since for some CRS data, we can observe sudden large jumps in a short time. For instance, with the exception of Estonia and Greece, the confidence intervals contain most of the data at the beginning of the period and some data points lie outside but close to the confidence interval (see Fig. 2). As it can be noted during the week between January 30th, 2012 and February 6th, 2012, there is a jump of 2–4 CRS points (due to the tensions derived from the negotiation of Greece's second rescue package and the time running out on this Greece rescue plan to bring Greece's debt back to a sustainable level). Furthermore, the case of Greece is well approximated since it is experiencing a dramatic drop because of its uncertain political situation at the current moment. The worst approximation corresponds to Estonia. It can be explained because the Estonia's economy has been the fastest growing economy in the European Union which has been forecast to have grown by 8 % during 2011 (that is more that 5 times faster growth than the European Union as a whole); however, with the Eurozone crisis still unresolved, this high growth has started to slow down. Despite all these facts and in all cases, the trends are well depicted by the model. Except for Estonia, where there is a certain stability, there are decreasing trends in the CRS for the next year. Considering the uncertainty of the data, it is complicated to provide accurate approximations. In the case of some countries, this leads to predictions with 90 % confidence intervals larger than expected. However, it can be explained because of the jumps of 4 CRS points within a week (January 30th, 2012 to February 6th, 2012).

4 Conclusions

Convergence in Europe has slowed down and, in the worst case, even gone into reverse, especially in Southern European Countries. Country Risk Score (CRS) represents the level of confidence on each country and a measure of its economic health.

Core eurozone sovereigns have seen their scores deteriorate sharply since the global financial crisis started.

In this work, we present a diffusion model to study the dynamics of the Country Risk Score (CRS), for a total of 27 European countries, which considers both the endogenous effect of each country politics and the contagion effect among them. Using data of CRS, we fit the model with the data estimating unknown autonomous behavior and transmission parameters. Then, we use an adapted cross-validation technique in order to provide probabilistic predictions over the next year (April 2012 until March 2013) taking into account that most of CRS data should be inside the confidence intervals corresponding to their time instants.

The obtained results depict quite well the evolution of the CRS for most of the countries, despite the jumps and uncertainty in the data within some periods. It can be observed a decreasing trend in the CRS until March 2013, except in Estonia where there is a certain stability. High unemployment rates and the loss of confidence in sovereigns bonds are not helping to reverse this trend. Moreover, it seems like the adopted measures in order to face these problems, austerity plans and price stability in the Eurozone as a whole, are squeezing the peripheral economies and will slow down the productivity growth.

As we have remarked, it should be pointed out that mathematical modelling with probabilistic predictions may be a powerful tool where policy makers are able to design strategies, simulate different situations and analyse the effect of changes. Among these strategies and despite the resistance of some Northern and Continental European Countries, a growing trend argues that one of the possible solutions must entail loosen monetary conditions by cutting interest rates and printing money to buy bonds, on one hand, or, on the other hand and in the long-term, the well known Eurobonds or some form of joint liability for countries debts might also be considered.

References

1. Ahluwalia, P.: Discriminating contagion: an alternative explanation of contagious currency crises in emerging markets. IMF Working Paper WP/00/14. http://http://www.worldcat.org/title/discriminating-contagion-an-alternative-explanation-of-contagious-currency-crises-in-emerging-markets/oclc/43515556 (2000). Accessed 15 May 2012
2. Bayoumi, T., Fazio, G., Kumar, M., MacDonald, R.: Fatal attraction: using distance to measure contagion in good times as well as bad. Rev. Financ. Econ. **16**, 259–273 (2007)
3. Brauer, F., Castillo-Chávez, C.: Mathematical Models in Population Biology and Epidemiology. Springer, New York (2001)
4. http://www.euromoney.com/poll/10683/PollsAndAwards/Country-Risk.html. Accessed 15 May 2012
5. Dornbusch, R., Park, Y.C., Claessens, S.: Contagion: understanding how it spreads. World Bank Res. Obs. **15**(2), 177–197 (2000)
6. Dungey, M., Fry, R., González-Hermosillo, B., Martin, V.L.: Empirical modelling of contagion: a review of methodologies. Quant. Finance **5**(1), 9–24 (2005)
7. Edwards, S., Susmel, R.: Interest rate volatility and contagion in emerging markets: evidence from the 1990. NBER working paper series 7813. http://www.nber.org/papers/w7813 (2000). Accessed 15 May 2012

8. Edwards, S., Susmel, R.: Volatility dependence and contagion in emerging equity markets. NBER working paper series 8506. http://www.nber.org/papers/w8506 (2001). Accessed 15 May 2012
9. Eichengreen, B., Rose, A., Wyplosz, C.: Contagious currency crises. NBER working paper series W5681. http://www.nber.org/papers/w5681 (1996). Accessed 15 May 2012
10. Fernández-Arias, E., Montiel, P.J.: The surge of capital inflows to developing countries: an analytical overview. World Bank Econ. Rev. **10**(1), 51–77 (1996)
11. Forbes, K., Rigobon, R.: Measuring contagion: conceptual and empirical issues. In: Claessens, S., Forbes, K.J. (eds.) International Financial Crises, pp. 43–66. Kluwer, Boston (2001)
12. Frambach, R.T.: An integrated model of organizational adoption and diffusion of innovations. Eur. J. Mark. **27**, 22–41 (1993)
13. Hamerly, G., Elkan, C.: Alternatives to the k-means algorithm that find better clusterings. In: Proceedings of the 11th ACM International Conference on Information and Knowledge Management, pp. 600–607 (2002)
14. Hoti, S., McAleer, M.: An empirical assessment of country risk rating and associated models. J. Econ. Surv. **18**(4), 540–588 (2004)
15. Kodres, L., Pritsker, M.: A rational expectations model of financial contagion. J. Finance **57**(2), 769–799 (2002)
16. MacKay, D.: Information Theory, Inference and Learning Algorithms, pp. 284–292. Cambridge University Press, Cambridge (2003)
17. Mahajan, V., Muller, E., Bass, F.M.: New product diffusion models in marketing: a review and directions for research. J. Mark. **54**, 1–26 (1990)
18. Masson, P.R.: Contagion: monsoonal effects, spillovers and jumps between multiple equilibria. In: Agenor, P., Miller, M., Vines, D. (eds.) The Asian Crises: Causes, Contagion and Consequences, pp. 265–283. Cambridge University Press, Cambridge (1999)
19. Mathematica. http://www.wolfram.com/products/mathematica
20. Nelder, J.A., Mead, R.: A simplex method for function minimization. Comput. J. **7**, 308–313 (1964)
21. Sachs, J., Tornell, A., Velasco, A.: Financial crises in emerging markets: the lessons from 1995. NBER working paper series 5576. http://www.nber.org/papers/w5576 (1996). Accessed 15 May 2012
22. Taylor, M.P., Sarno, L.: Capital flows to developing countries: long-and short-term determinants. World Bank Econ. Rev. **11**(3), 451–470 (1997)
23. Zhang, D., Ntoko A.: Mathematical model of technology diffusion in developing countries. In: Computational Methods in Decision-Making, Economics and Finance, pp. 526–539. Kluwer Academic Publisher, Boston (2002)

Chapter 5
Determining the 'Fault Zone' of Fall Events in Long Term Care

Vijay Kumar Mago, Ryan Woolrych, Stephen N. Robinovitch and Andrew Sixsmith

Abstract ***Background:*** Fall incidents in long-term care facilities are complex events, involving an interplay of factors at the individual and organisational level. However, research and care interventions have primarily focussed on examining specific intrinsic and extrinsic risk factors, rather than capturing a holistic understanding of the fall event. As a result, there is a paucity of research which has captured the contributory factors of falls at the personal, interpersonal and organisational level. Such understandings are important for broadening the evidence base and developing effective interventions which address potential causes at all levels. ***Methods:*** An ecological, systems based approach to analysing falls in long-term care allows a more holistic and systemic understanding of the fall event to emerge. This chapter applies a fuzzy cognitive map (FCM) modelling approach to understanding falls within long-term care, using an ecological analysis to demonstrate the inter-relationships between the various contributory factors at the micro individual level. The model is driven by case study data collected of a fall incident taking place in a long-term care facility in Greater Vancouver. ***Results:*** For experimentation, various real-life scenarios have been created to test the model. The results demonstrate that FCM models provide a potentially valuable tool for conceptualising the complexity of falls within long-term

V. K. Mago (✉)
Department of Computer Science, University of Memphis, Memphis, TN, USA
e-mail: vkmago@memphis.edu

R. Woolrych
Gerontology Research Centre, Simon Fraser University, Harbour Centre, Vancouver, Canada
e-mail: rwoolryc@sfu.ca

A. Sixsmith
Department of Gerontology, Simon Fraser University, Harbour Centre, Vancouver, Canada
e-mail: ajs16@sfu.ca

S. N. Robinovitch
Department of Biomedical Physiology and Kinesiology, Simon Fraser University, Burnaby, Canada
e-mail: stever@sfu.ca

V. K. Mago and V. Dabbaghian (eds.), *Computational Models of Complex Systems*, Intelligent Systems Reference Library 53, DOI: 10.1007/978-3-319-01285-8_5,

© Springer International Publishing Switzerland 2014

care, providing a means for better educating healthcare practitioners and designing targeted interventions. ***Conclusions:*** The ecological model provides the framework for analysing falls at various levels within the context of the long-term care facility and beyond (e.g., culture of falls prevention and falls prevention policy). The FCM technique provides a tool for visually conceptualising this complexity, highlighting the inter-relationships between individual factors and presenting the results in such a way that can be easily interpreted by academicians, practitioners and support staff.

1 Introduction

Falls are the number one cause of unintentional injury among older adults in Canada, including over 90 % of hip fractures and wrist fractures, and a large percentage of head and spine injuries [37]. According to the Canadian Institute for Health Information [35], falls are the second leading cause of injury-related hospitalization. In the USA, the incidences of unintentional fall injuries led to the death of 19,700 people in 2008 whilst the death rate among older adults due to falls has seen a sharp rise over the past decade [10]. The Public Health Agency of Canada has put forward a number of recommendations to help prevent falls and injuries by minimizing the risks [34], but fall incidences continue to represent a significant care and cost burden [14] with the length of stay in hospital being twice as long in each age group than for all causes of hospitalization for people over the age of 65, and representing an annual cost of 2.8 billion in 2004 [29]. Moreover, falls have considerable social cost to the individual, leading to post-fall restrictions on mobility that have a significant impact on self-reported independence, autonomy and quality of life [32].

In understanding falls amongst older people, the focus of much of the research thus far has been on understanding and preventing falls amongst community-dwelling adults [13]. This is perhaps surprising given that residents in long-term care (LTC) are up to twice as more likely to experience both severe injury as a result of a fall and the likelihood of a recurrent fall than community dwelling older adults [7]. Research indicates that up to 50 % of residents within long-term care will fall each year and that up to 40 % of these will fall more than once. Moreover, hip fractures and other major injuries are particularly common as a result of falls within residential care facilities [26, 30] with fall incidents representing the primary cause of injury in older adults [23].

The research that has been conducted within long-term care facilities has focussed primarily on establishing the intrinsic and extrinsic factors associated with falls [33]. Intrinsic factors include poor gait and imbalance, muscle weakness, vision impairment. Extrinsic factors include immediate environmental hazards, poor lighting and the availability and use of assistive aides. While this research has helped in understanding falls in long-term care, it primarily focusses on principal risk factors and establishing cause and effect relationships,which does not accurately capture the inter-related ways in which these factors come together at the time of the fall. Moreover, what we know about falls in long-term care is predicated upon post-fall accounts of incidents, collected either through existing reporting mechanisms or post-fall witness observations [11]. Whilst these data sources provide useful information, there are weaknesses in their reliability, susceptible to memory recall bias and prescriptive forms which do not provide an accurate narrative of the fall event. To address these weaknesses, [31] utilised real video of fall incidents within long-term care to analyse the biomechanical causes of falls within long-term care.

The design of falls interventions in long-term care have primarily been focussed on addressing factors in isolation, for example, exercise regimes to improve muscle strength. Whilst multi-factorial fall interventions have typically looked to combine the different risk factors, for example, exercise, medication and environmental factors, this has not addressed some of the broader contributory factors of falls at the interpersonal and organisational level. This is indicative of the person-centred and reactive approach taken to identifying levels of risk in the care facility, for example, through the application of individual fall risk assessments and care plans. More recently, researchers have suggested technology based solutions such as electronic sensors and equipment including fall detectors, door monitors, bed alerts and pressure mats [4]. Miskelly [24] conducted a study that assessed the capabilities of an accelerometer in the iPhone to identify changes in gait when walking on a flat surface. Chan et al. [9] utilised ultra-wideband radio real-time location sensor network to record variability in voluntary movement paths of assisted living facility residents. Similarly, Kearns et al. [16] applied fractal analysis on path variability to predict fall events among residents in assisted living, whereby greater variation from a straight line or path increased the likelihood of falling. The aim of these various technologies has been to either alert caregivers to an emergency in the event of a fall or to identify changes in patterns of everyday behaviour. Whilst these interventions have been useful in determining the movements of older people, they have not considered how these interventions are part of the broader social, cultural and organisational context of care settings, thereby addressing isolated components of the problem. What is needed is an approach to conceptualising falls as holistic events, as an outcome of a number of inter-related factors across all levels.

Literature has highlighted the need for research which examines the causes of falls and the circumstances surrounding these adverse events [6], for example, environmental indicators of falls have primarily been approached from retrospective accounts which have identified trips and slips resulting from environmental hazards. There is a paucity of research examining how social, behavioural and situational factors combine prior to the fall incident, although research has identified the importance of these factors in isolation [30, 33]. Moreover, LTC is a challenging work and care environment, delivering care to residents who have complex care requirements and a broad range of behaviours associated with old age. In this sense, the LTC setting can be seen as a complex system, involving many linkages, inter-relationships and dynamics at the organisational level which need to be understood [1], for example, the recognition that resident well-being in LTC is influenced by the delivery of one-to-one care which is constrained by staffing levels and decisions made at the higher level. Thus, fall incidents need to be understood as an outcome of a complex interplay of factors operating at different contextual levels within a LTC facility. This perspective has been adopted by Zecevic and colleagues [41] who describe fall incidents through a systems based approach. A fall is seen as a result of a chain of events, within which a number of failures or errors in the system combine and contribute to the fall incident. Zecevic et. al. suggests that these system failures can occur at the organisational level (such as organisational culture and policy), through supervision (such as monitoring and observation), stemming from preconditions (such as poor equipment design or medication) and as a result of unsafe acts and decisions (such as walking and turning). This approach identifies the need for a holistic approach to falls prevention, recognising that there is a multiplicity of precipitating factors with interactions between them. Conceptualising this complexity is challenging, yet important if we are to translate knowledge effectively amongst healthcare practitioners to better design interventions.

The main goal of the research presented in this chapter is to develop a simulation model of falls in LTC facilities that will incorporate, simultaneously, the multiplicity of factors contributing to the fall incident at the micro level. This is one layer of the ecological framework proposed by Bronfronbrenner and describes the individual characteristics of the faller and their immediate surroundings. There is the potential to apply the tool described in this chapter to other levels (meso, exo, macro and chromosphere). The **meso level** defines the social system or the interpersonal factors, the **exo level** identifies the prevailing organizational structure of LTC and the **macro level** refers to the cultural values and broader policies in place at wider institutional level. The **chronosphere** is the temporal frame for the fall event, for example, factors related to personal histories or patterns in falls over time. More details can be found in Sect. 2.2. The model has the potential to represent the effect of the interaction between individual factors and the impact of these interactions on the possibility of a fall incident occurring across all these levels.

The model presented in this chapter will be driven by data collected in a long-term care facility in Metro Vancouver as part of the Technology for Injury Prevention in Seniors (TIPS) research project, funded by the Canadian Institutes of Health Research, the aim of which is to better understandings of falls in long-term care and to design effective interventions. The research presented in this chapter is part of a sub-project of the TIPS program tasked with examining the environmental factors of falls within long-term care and the broader context within which falls occur. Given the complexity of falls within LTC, a case study approach was adopted to the research, with each case representing a fall incident occurring within a long-term care facility in greater Vancouver, BC, Canada. A mixed methods approach was undertaken to case study development, incorporating primary (interviews and focus groups with fallers, nurses, care aides and management) and secondary data (fall incident reports, resident medical notes, video of fall incidents). Whilst case studies are necessary to examine the fall incident through the eyes of multiple actors involved in the fall, they generate substantial amounts of data which need to be accurately presented in such a way that they are interpretable by others whereby better understandings and interventions can emerge. To this end, this chapter combines a fuzzy cognitive mapping technique with an ecological understanding of a fall incident, drawing upon actual case study data to better conceptualise the complexity of falls in LTC.

2 Methodological Framework

2.1 Fuzzy Cognitive Map

Fuzzy cognitive map was first introduced by Bart Kosko as a causal reasoning method [18]. The main advantage of the FCM approach is that it allows fuzzy causality among domain concepts and the network structure permits the possibility of causal propagation. This is particularly relevant to falls research as the information regarding fall incidents is not crisp or exact in nature. Moreover, due to the large number of causal factors, it becomes highly complex to conceptualise. That is why researchers have only succeeded in defining the causal relationship between a limited number of concepts in any single research work. For instance, authors of [36] could only suggest that older people with cognitive impairment and dementia are more at risk of a fall, but could not elaborate further on other related aspects. Moreover, we assume that the degree of causality suggested by these works cannot be generalized. Similarly, authors of [15] investigated fall events among frail older adults living in LTC and highlighted the causal relationship between frailty and an increase in falls, noting that, "The overall frequency of falls related to acute diseases in the residential care facilities is most probably related to the frailty of the residents". Even though the knowledge about the causality exists in the literature and the researchers are aware of the various primary causes of falls within long-term care, analyzing all factors in a single snapshot is challenging and requires innovative approaches. FCM provides a platform to address this challenge. Interested readers can refer to the recent works of the authors [12, 20–22, 28].

Formally, FCM is a collection of nodes, say n. These nodes represent concepts of the domain. Nodes are connected with directed edges and on each edge there is a weight which can be either positive or negative. The type of the weight signifies the form of correlation between two concepts, i.e., positive or negative. Assume that there are two concept nodes C_i and C_j which are connected and the weight is w_{ij}. One can construct an adjacency matrix using all the weights. To run a simulation, an initial vector is supplied to the following mathematical formula

$$A_i^{\tau+1} = f(A_i^{\tau} + \sum_{i}^{n} A_j \times w_{ij}) \tag{1}$$

where A_i and A_j are the initial values of concepts C_i and C_j respectively and τ represents the iteration count. Other representation suggested in [8] is:

$$[A_1, \ldots, A_n]^{\tau+1} = f\left([A_1, \ldots, A_n]^{\tau} \times \begin{bmatrix} 0 & \cdots & w_{1n} \\ \vdots & \ddots & \vdots \\ w_{n1} & \cdots & 0 \end{bmatrix}\right) \tag{2}$$

$$f(x) = \frac{1}{1 + e^{-\lambda x}} \tag{3}$$

where λ value decides how fast the system converge.

The threshold function 'f' is used to control the values of the concepts between -1 and $+1$. The most common threshold functions are sigmoid function, hyperbolic tangent function, step function, and linear threshold function. More recently authors of [19] suggested another function, sinusoidal-type, and argued that their proposed function is more suitable for inferential process for clinical decision making. Interested readers may read [38] for comparing the inference capabilities of different threshold functions. In this research we used the most common threshold function which is defined in Eq. 3. The FCM model at the *micro level* is shown in Fig. 1.

2.2 Ecological Model

The long-term care environment can be particularly hazardous for older adults who experience functional impairment or limitations in their physical mobility. In ambulating around the home or care setting such restrictions bring about the increase likelihood of falling through personal factors (pre-existing conditions, behaviours), interpersonal (interactions with care staff, other residents and the immediate environment), situational (time of day, length of stay) and broader organisational factors (staffing level). This complexity is further deepened by broader macro factors related to those influences outside of the organisation, for example, those indicators that impact on the culture of falls prevention such as funding mechanisms, falls prevention policy and cultures of care delivery.

To model this multi-layered complexity we can utilise Bronfronbrenner's Ecological Systems Model. Bronfenbrenner developed an ecological systems approach to define and understand human development within the context of the system of relationships that form the person's environment. He defined the ecological systems approach as:

"*The ecology of human development is the scientific study of the progressive, mutual accommodation throughout the life course between an active, growing human being and the changing properties of the immediate settings in which the developing person lives. [This] process is affected by the relations between these settings and by the larger contexts in which the settings are embedded* [3]".

The approach can be extended across different environments and organisations and is valuable in describing complex systems such as fall events [27]. The model has the potential to articulate the complexity of falls within long-term care, incorporating an understanding of the fall as a system of relationships where the fall can be seen as resulting from inter-related factors within the environment. According to Bronfenbrenner's initial theory (1989), the environment, is comprised of four layers of systems which interact in complex ways. Changes or conflict in any one layer will ripple and trickle down to other layers. To study a fall, we must look not only at the

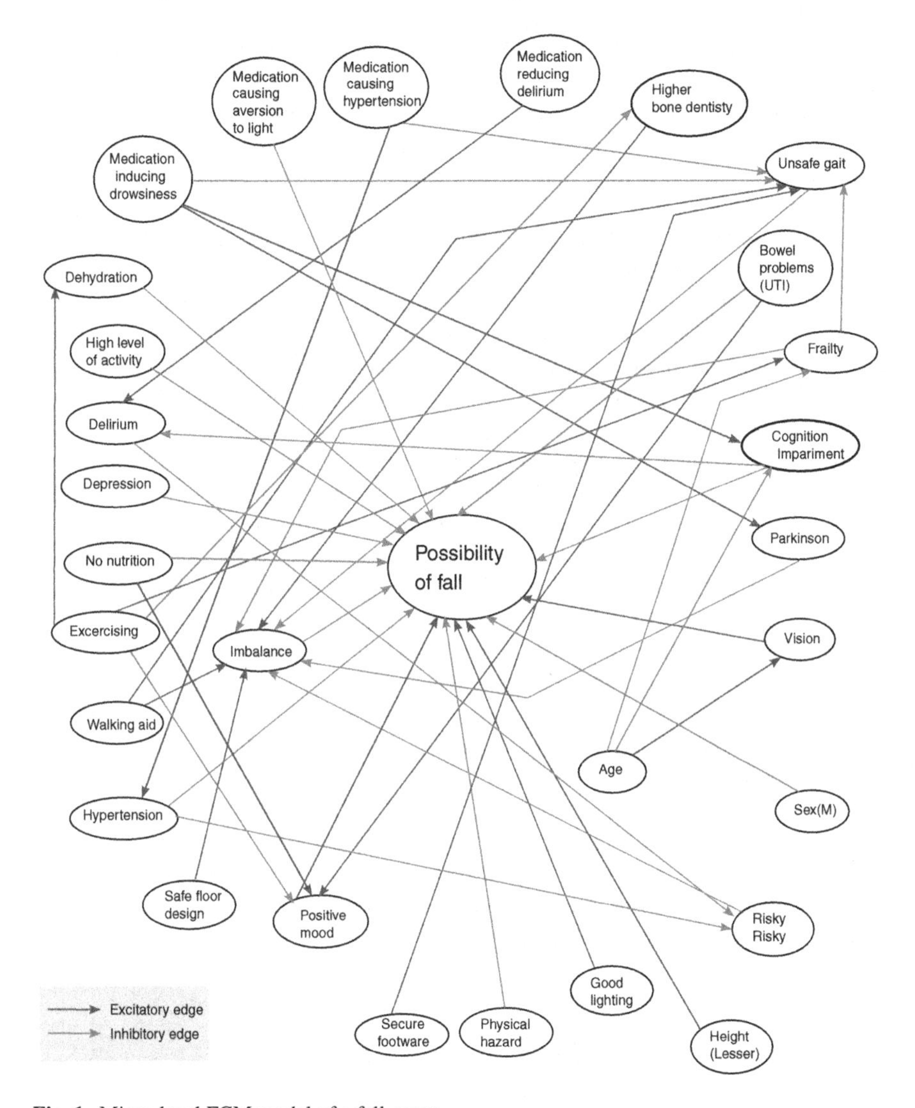

Fig. 1 Micro-level FCM model of a fall event

faller and their immediate environment, but also at the influence at the contextual and situational level as well. The layers contained within the ecological systems model are the microsystem, the mesosystem, the exosystem, the macrosystem and the chronosystem:

Microsystem: this is the layer that would be closest to the individual and includes interpersonal factors that may have contributed to the fall. The influences between the person and these structures are seen as bidirectional i.e., the person influences and is influenced by the microsystem. Within the context of falls, the microsystem encompasses the individual characteristics of the faller (for example, dementia and preconditions) and the relationships and interactions a faller has with their immediate surroundings (for example, medication and environmental awareness) [2].

Mesosystem: this layer provides the connection between the structures of the fallers microsystem [2], for example, that encompassing the social system or the interpersonal factors. An example of the mesosystem can be seen in the interactions and dynamics between two of its microsystems, for example, the interaction between the faller and the care aide.

Exosystem: this layer defines the larger social system in which the faller does not function directly. The structures in this layer impact the faller by interacting with some structure in his/her microsystem [2]. The exosystem encompasses events, contingencies, decisions, and policies over which the developing person has no influence. The faller may not be directly involved at this level, but he does feel the positive or negative force involved with the interaction with his/her own system. Within the context of falls, this places the aging person in the context of the LTC as a setting, influenced by factors such as carer availability, carer competence etc.

Macrosystem: this layer may be considered the outermost layer to the faller. While not being a specific framework, this layer is comprised of cultural values, customs, and laws [2]. The effects of larger principles defined by the macrosystem have a cascading influence throughout the interactions of all other layers. For example, the way in which care is delivered, funding from healthcare agencies, cultural approaches to falls. This, in turn, affects the structure in which the carer functions. The carers ability or inability to carry out that responsibility toward the faller within the context of the fallers microsystem is likewise affected.

Chronosystem: this system encompasses the dimension of time i.e., providing a temporal component, as it relates to the fallers environment. Elements within this system can be either external, such as changes to the environment or relocation to another long-term care setting, or internal, such as the physiological changes that occur with the aging of an older person.

2.3 Case Studies of Fall Incidents

The FCM model is driven by data generated by a case study conducted in a long-term care facility in greater Vancouver, using qualitative data generated by interviews and focus groups with care aides, nurses, care managers and directors and information generated from the case notes and medical notes held at the long-term care facility.

The qualitative data was thematically analysed to identify recurring themes and patterns from the data. These recurring themes and patterns highlighted a number of factors at the micro, meso and macro level which contributed to the fall event. A number of the factors at the micro level were extrapolated and are discussed in more detail as cases in Sect. 3.

3 Experimental Results

To conduct different simulation experiments, a software system was implemented using MATLB 7.9.0 (R2009b). The system produces the visualization of the sample cases for validation. The validation process depends on the face validity [25] of the simulation results produced by the system. Face validity refers to the intentional measurement. For instance, if a set of experts assume that the resident may experience a fall, and the system also predicts the possibility of a fall, then the system is validated. With the help of experts the authors constructed various plausible test cases which are discussed in the remaining section.

3.1 Case 1: Role of Medicine in Fall Events

In this case example, we assume that the resident is on a number of different types of medication to control pre-existing conditions including hypertension, delirium and other age related problems. These medicines may also cause aversion to light or drowsiness which in themselves represent specific risk factors in falls and can impact on gait and balance. Authors of [5] highlighted that, on average, residents in LTC are administered 4.5 prescription drugs and 2 over the counter medicines every day, and that these drugs have potential negative side effects. Given these circumstances, one would assume that the resident would have an increased likelihood of falling, but if you notice that the simulation result shown in Fig. 2, the resident under observation will have unstable gait (pink line), and initially experience some kind of imbalance (somewhat pink), but the side effects of medication alone will not contribute to the fall. This finding is supported by the literature which identifies that modifying medication can reduce the risk of falling but not remove them completely [39].

3.2 Role of the Physical Environment in LTC

In this case the authors assume that the resident is experiencing an unsafe gait and that this risk factor is known to the care providers. As a preventive measure, the care staff have provided a supportive environment including good lighting, suitable floor design, safe footwear and a walking aid (which the resident uses all of the time). Nurses have also ensured that any potential environmental hazards in the immediate surroundings have been identified and removed. Given these interventions, one would assume that the resident is less likely to experience a fall. Our simulation result also predicts the same outcome in Fig. 3. The resident who is using a walking aid to navigate around the walking environment and has supportive aides within the environment, is less likely to experience a fall.

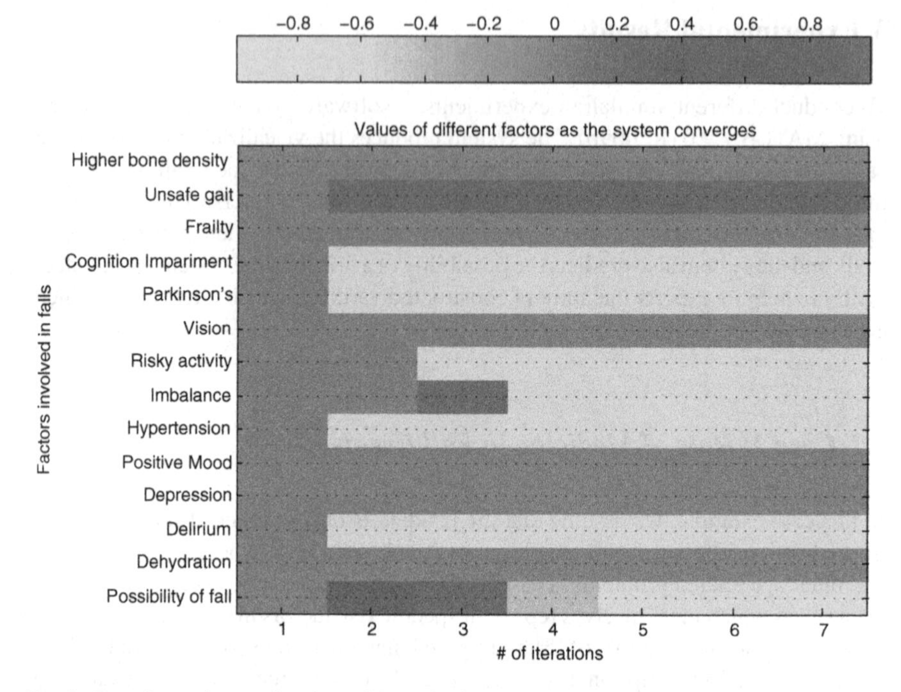

Fig. 2 Experimental scenario of a resident who is on medication. The system stabilizes after a few iterations with the result that the resident will have an *Unsafe gait* (*pink colored bar*) and the possibility of a fall being negative (-1, represented in *light blue color*)

Types of fall prevention measures have been well documented by the Public Health Agency of Canada [40] and care providers are trained to perform these fall preventive strategies. Our simulation endorses the fact that if all preventive strategies are implemented fully, i.e., the initial state of these concepts in the model are $+1$, then the resident, even with unsafe gait, will not fall as interventions are in place to compensate for these safety gaps. Practical implementation of these strategies is complex and sometimes the results may vary given the unpredictability of the care delivery setting [17]. However, these are the assumptions generated by the data and the results produced by the system are coherent.

3.3 Case 3: Uncertain Case Where Multiplicity of Factors are Present

We assume a female resident who is 70 years old is experiencing a number of health conditions that increase the likelihood of a fall whilst also exhibiting factors that might reduce the probability of a fall. For instance, the resident is mildly cognitively impaired, experiences hypertension, imbalance, delirium and dehydration. Yet the

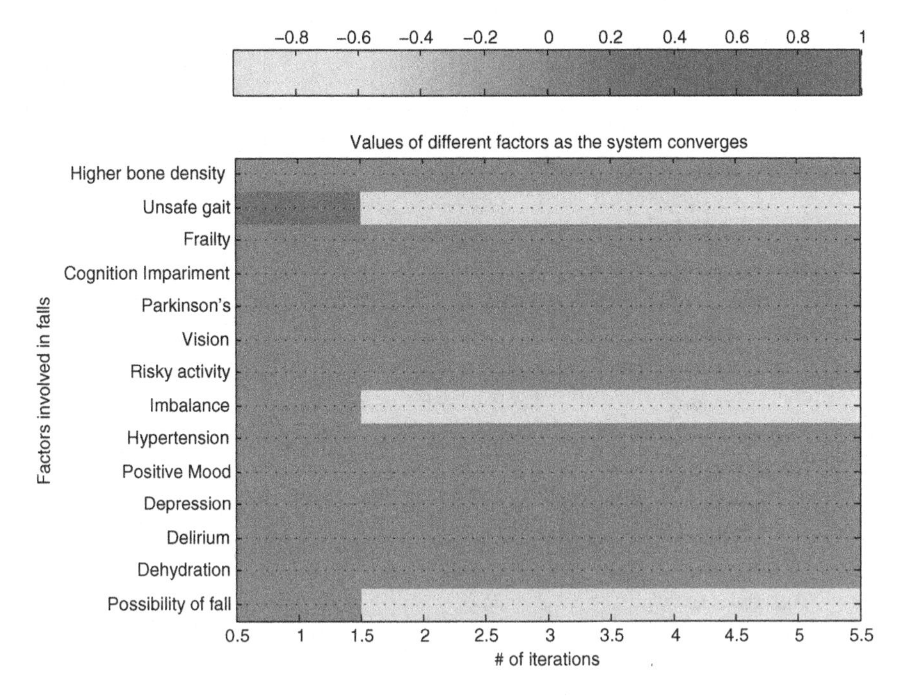

Fig. 3 Experimental scenario of a resident who has an unsafe gait

resident has a very strong body frame and bone density. In addition to intrinsic factors, the environmental layout of the LTC features some factors that contribute to falls and others that prevent the likelihood of a fall. For example factors such as poor lighting and the presence of physical hazards in the bedroom increases the likelihood of falling, but the availability of grab bars and a stable floor design may reduce the likelihood of a fall. Given the combination of positive and negative influences exerted by intrinsic (health status) and extrinsic (environmental) issues, one would be unsure of the outcome. When we run the simulation, the negative influences outweigh the positive influences and this is shown in Fig. 4. The system predicted that the possibility of a fall is high for this resident. This is because of the reason that negative forces (unsafe gait, frailty, cognition impairment, imbalance, etc.) are exerting pressure on the positive factors.

The most important and interesting aspect of Fig. 4 is the number of iterations the system took to stabilize, at more than 40. If we compare the two previous cases, it took approximately five times the iterations to converge to a solution where the possibility of a fall is determined to be very high. Clearly the system took more computational processing to predict the result as the interplay between positive and negative forces created a complex situation.

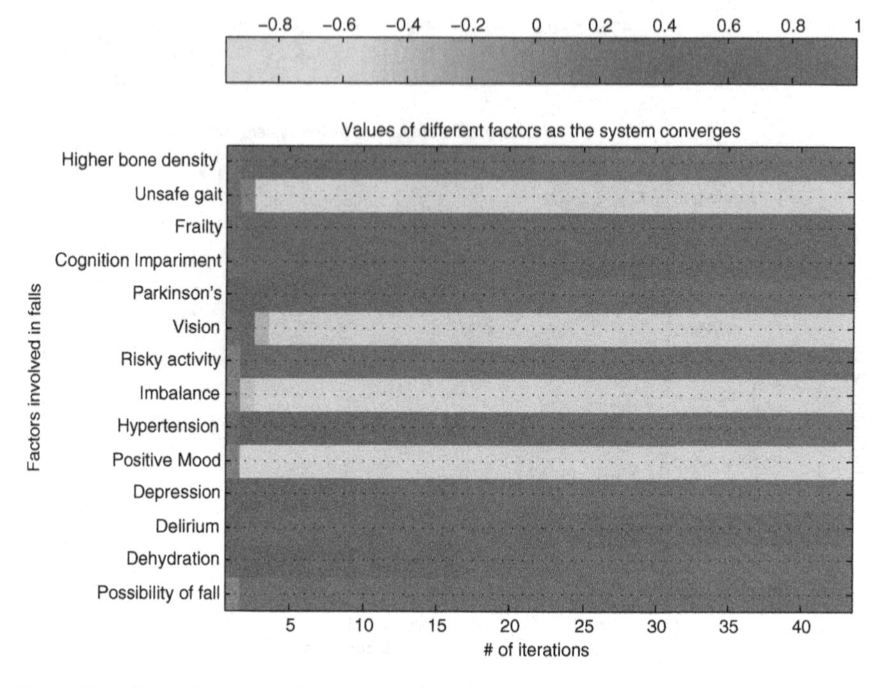

Fig. 4 Experimental scenario with numerous factors are influencing together

4 Discussion and Future Work

The ecological model provides the framework for analysing falls at various levels within the context of LTC and beyond to include various social and cultural components that ultimately have an impact on fall incidents. The FCM technique provides a tool for conceptualising this complexity, highlighting the inter-relationships between individual factors and presenting this in such a way that it can be understood by academics, practitioners and frontline workers. In this sense, the ecological model provides the base for developing FCM at various levels of abstraction. Using expert opinion collected from the field, the model has the potential to be rooted in the experiences of older people themselves and those delivering care to the residents within LTC. If it is the task of the researcher to supply decision makers (e.g. people who are responsible for the care and support of older people) with coherent and usable information about some area of concern, then it is important that all relevant issues are incorporated within the content universe. There is a tendency to simplify phenomena for various reasons. For example, the need for scientific "rigor" may actually mean that many important issues are excluded from user research, because they are not amenable to measurement. There is also a tendency to conceive phenomena as "problems" to facilitate practical intervention (e.g. technologies or services) in the way that reinforces conventional practice and received wisdom. This harks back to the earlier discussion of "needs ", where the conceptualization of the problem may be based more on the requirements of the service provider than on the user. While this may be helpful in improving the operations of existing systems and services, it fails to address alternative approaches and new ideas about the help and support provided to older people. However, there is an equal problem of making something too complex and all-embracing, where the complexity ultimately undermines any practical action. The "trick", therefore, is to define any theoretical framework in a way that preserves the integrity of the phenomena in question, while at the same time presenting these ideas so that they are understandable and accessible.

There are a number of limitations with the research described here. Firstly, this chapter presents only a subset of micro level factors in the case examples. There is a need to expand this to the meso and macro level to fully understand how contributory factors of falls cascade between layers. Secondly, whilst the model was driven by the opinions of experts collected through the case study work, the model has not been further validated by experts themselves. Engaging them in the model development would allow for further refinement of and iteration of the model. Thirdly, there is a need to integrate a temporal component into the model. This is known as the chronosphere in the ecological model and recognises that fall incidents are also deeply situated in the personal history of the older person, and influenced by factors involving ageing over time.

The potential outcome of this work is a tool for analyzing contributory factors of falls in LTC for better policy making. The approach has two key benefits in practical terms. Firstly, a comprehensive model of falls will allow policy-makers and practitioners to effectively develop and target interventions that consider the wide range of real-world factors. Secondly, the model will allow practitioners to estimate and compare the likely effects of different interventions and scenarios by using simulations prior to real-world interventions, thus reducing negative side effects in terms of risks to residents and service costs.

Acknowledgments We thank Gerontology Research Centre and MoCSSy program at SFU for providing financial support for the project. The authors are also grateful to the IRMACS Centre, Simon Fraser University, BC, Canada for the technical support.

References

1. Anderson, R.A., Issel, L.M.: Nursing homes as complex adaptive systems: relationship between management practice and resident outcomes. Nurs. Res. **52**(1), 12 (2003)
2. Berk, L.E.: Child Development, 5th edn. Allyn & Bacon, Needham Heights (2000)
3. Bronfenbrenner, U.: Ecology of the family as a context for human development: research perspectives. Dev. Psychol. **22**(6), 723–725 (1986)
4. Browne, J., Covington, B., Davila, Y.: Using information technology to assist in redesign of a fall prevention program. J. Nurs. Care Qual. **19**(3), 218–225 (2004)
5. Cameron, K.: The role of medication modification in fall prevention. NCOA falls free: Promoting a national falls prevention action plan: Research review papers pp. 29–39 (2005)

6. Campbell, A.J., Borrie, M.J., Spears, G.F., Jackson, S.L., Brown, J.S., Fitzgerald, J.L.: Circumstances and consequences of falls experienced by a community population 70 years and over during a prospective study. Age Ageing **19**(2), 136–141 (1990)
7. Campbell, A., Robertson, M., Gardner, M., Norton, R., Buchner, D., et al.: Psychotropic medication withdrawal and a home-based exercise program to prevent falls: a randomized, controlled trial. J. Am. Geriatr. Soc. **47**(7), 850 (1999)
8. Carvalho, J.P.: On the semantics and the use of fuzzy cognitive maps and dynamic cognitive maps in social sciences. Fuzzy Sets Syst. **214**(0), 6–19 (2013) (Soft Computing in the Humanities and Social Sciences)
9. Chan, H., Zheng, H., Wang, H., Gawley, R., Yang, M., Sterritt, R.: Feasibility study on iphone accelerometer for gait detection. In: Pervasive Computing Technologies for Healthcare (PervasiveHealth), 2011 5th International Conference on, pp. 184–187. IEEE (2011)
10. Falls among older adults: An overview. Technical report, Centers for Disease Control and Prevention, National Center for Injury Prevention and Control, Division of Unintentional Injury Prevention (2012)
11. Feldman, F., Robinovitch, S.N.: Reducing hip fracture risk during sideways falls: evidence in young adults of the protective effects of impact to the hands and stepping. J. Biomech. **40**(12), 2612–2618 (2007)
12. Giabbanelli, P.J., Torsney-Weir, T., Mago, V.K.: A fuzzy cognitive map of the psychosocial determinants of obesity. Appl. Soft. Comput. **12**(12), 3711–3724 (2012)
13. Gillespie, L., Gillespie, W., Robertson, M., Lamb, S., Cumming, R., Rowe B.: Interventions for preventing falls in elderly people. Cochrane Database Syst. Rev. **15**(2), CD000340 (2009)
14. Heinrich, S., Rapp, K., Rissmann, U., Becker, C., König, H.H.: Cost of falls in old age: a systematic review. Osteoporos. Int. **21**(6), 891–902 (2010)
15. Jensen, J., Lundin-Olsson, L., Nyberg, L., Gustafson, Y.: Falls among frail older people in residential care. Scandinavian J. Public Health **30**(1), 54–61 (2002)
16. Kearns, W., Fozard, J., Becker, M., Jasiewicz, J., Craighead, J., Holtsclaw, L., Dion, C.: Path tortuosity in everyday movements of elderly persons increases fall prediction beyond knowledge of fall history, medication use, and standardized gait and balance assessments. J. Am. Med. Dir. Assoc. **13**, 665.e7 (2012)
17. Kerse, N., Butler, M., Robinson, E., Todd, M.: Fall prevention in residential care: a cluster, randomized, controlled trial. J. Am. Geriatr. Soc. **52**(4), 524–531 (2004)
18. Kosko, B.: Fuzzy cognitive maps. Int. J. Man Mach. Stud. **24**(1), 65–75 (1986)
19. Lee, I., Kim, H., Cho, H.: Design of activation functions for inference of fuzzy cognitive maps: application to clinical decision making in diagnosis of pulmonary infection. Healthc. Inform. Res. **18**(2), 105–114 (2012)
20. Mago, V., Woolrych, R., Sixsmith, A.: Understanding fall events in long term care using fuzzy cognitive map. Gerontechnology **11**(2), 343 (2012)
21. Mago, V.K., Bakker, L., Papageorgiou, E.I., Alimadad, A., Borwein, P., Dabbaghian, V.: Fuzzy cognitive maps and cellular automata: an evolutionary approach for social systems modelling. Appl. Soft. Comput. **12**(12), 3771–3784 (2012)
22. Mago, V.K., Mehta, R., Woolrych, R., Papageorgiou, E.I.: Supporting meningitis diagnosis amongst infants and children through the use of fuzzy cognitive mapping. BMC Med. Inform. Decis. Mak. **12**(1), 98 (2012)
23. Marks, R., Allegrante, J.P., Ronald MacKenzie, C., Lane, J.M.: Hip fractures among the elderly: causes, consequences and control. Ageing Res. Rev. **2**(1), 57–93 (2003)
24. Miskelly, F.: Assistive technology in elderly care. Age Ageing **30**(6), 455–458 (2001)
25. Nevo, B.: Face validity revisited. J. Educ. Meas. **22**(4), 287–293 (2005)
26. Norton, R., Campbell, A.J., Reid, I.R., Butler, M., Currie, R., Robinson, E., Gray, H.: Residential status and risk of hip fracture. Age Ageing **28**(2), 135–139 (1999)
27. Nowak, A., Hubbard, R.E.: Falls and frailty: lessons from complex systems. JRSM **102**(3), 98–102 (2009)
28. Pratt, S.F., Giabbanelli, P.J., Jackson, P., Mago, V.K.: Rebel with many causes: A computational model of insurgency. In: Intelligence and Security Informatics (ISI), 2012 IEEE International Conference on, pp. 90–95. IEEE (2012)

29. Prevention of falls and injuries among the elderly. Technical report, A Special Report from the Office of the Provincial Health Officer, British Columbia, Canada (2004)
30. Ramnemark, A., Nyberg, L., Borssen, B., Olsson, T., Gustafson, Y.: Fractures after stroke. Osteoporos. Int. **8**(1), 92–95 (1998)
31. Robinovitch, S.N., Feldman, F., Yang, Y., Schonnop, R., Leung, P.M., Sarraf, T., Sims-Gould, J., Loughin, M.: Video capture of the circumstances of falls in elderly people residing in long-term care: an observational study. The Lancet **381**(9860), 47–54 (2013)
32. Rowe, S.M., Song, E.K., Kim, J.S., Lee, J.Y., Park, Y.B., Bae, B.H., Hur, C.I.: Rising incidence of hip fracture in gwangju city and chonnam province, korea. J. Korean Med. Sci. **20**(4), 655–658 (2005)
33. Rubenstein, L.Z., Josephson, K.R., et al.: Falls and their prevention in elderly people: what does the evidence show? Med. Clin. North Am. **90**(5), 807–824 (2006)
34. Seniors and aging—preventing falls in and around your home. Technical report, Health Canada (2006)
35. Series on seniors: Seniors and falls. Technical report, Canadain Institute of Health Information, Ottawa (2010)
36. Shaw, F., Bond, J., Richardson, D., Dawson, P., Steen, I., McKeith, I., Kenny, R.: Multifactorial intervention after a fall in older people with cognitive impairment and dementia presenting to the accident and emergency department: randomised controlled trial. Br. Med. J. **326**(7380), 73 (2003)
37. Sran, M., Stotz, P., Normandin, S., Robinovitch, S.: Age differences in energy absorption in the upper extremity during a descent movement: implications for arresting a fall. J. Gerontol. A Biol. Sci. Med. Sci. **65**(3), 312 (2010)
38. Tsadiras, A.: Comparing the inference capabilities of binary, trivalent and sigmoid fuzzy cognitive maps. Inform. Sci. **178**(20), 3880–3894 (2008)
39. Vu, M., Weintraub, N., Rubenstein, L.: Falls in the nursing home: are they preventable? J. Am. Med. Dir. Assoc. **5**(6), 401–406 (2004)
40. You can prevent falls! Technical report, Public Health Agency of Canada (2011)
41. Zecevic, A.A., Salmoni, A.W., Lewko, J.H., Vandervoort, A.A., Speechley, M.: Utilization of the seniors falls investigation methodology to identify system-wide causes of falls in community-dwelling seniors. The Gerontologist **49**(5), 685–696 (2009)

Chapter 6
Classification of Child Disability Using Artificial Neural Network

Jumi Kalita, Kandarpa Kumar Sarma and Pranita Sarmah

Abstract There are millions of children suffering from disability across the globe. Due to its implicit characteristics, the classification and diagnosis of children with disability has long been a difficult issue. This paper proposes a systematic approach for classification of potential cases more accurately and easily by use of Artificial Neural Network (ANN). In this paper, an attempt has been made to apply Artificial Neural Networks (ANN) for classification of disability in children based on the combination of symptoms. ANNs are nonparametric statistical tools designed using bio-inspired approaches. ANNs have been configured for application in different fields as diverse as engineering technology to financial forecasting. The primary inspiration of the working principle of the ANN is created by mimicking the functioning of the human brain. ANNs can learn input patterns and use this knowledge to make predictions. ANNs are known to be adaptive, robust and can learn process data irrespective of methods of creation. All these spin-offs can be associated with this approach of predicting child disabilities using ANN. Such an approach has been applied in this paper to classify the type of disability and relating them to the symptoms observed at birth of a child along with mother's health condition during pregnancy. The advantage of such a system is derived from the strengths of the ANN. The paper proposes a Multi Layer Perceptron (MLP)-based ANN model for classifying types of disability among children. The model comprises of a single input layer which correspond to different symptoms at birth of a baby and maternal health during pregnancy and

J. Kalita (✉)
Department of Statistics, Lalit Chandra Bharali College, Guwahati, Assam 781011, India
e-mail: jumikalita@yahoo.co.in

K. K. Sarma
Department of Electronics and Communication Technology, Gauhati University, Guwahati, Assam 7810 14, India
e-mail: kandarpaks@gmail.com

P. Sarmah
Department of Statistics, Gauhati University, Guwahati, Assam 7810 14, India
e-mail: pranitasarma@gmail.com

V. K. Mago and V. Dabbaghian (eds.), *Computational Models of Complex Systems*, Intelligent Systems Reference Library 53, DOI: 10.1007/978-3-319-01285-8_6,

© Springer International Publishing Switzerland 2014

outputs corresponding to different types of disability. The preliminary results obtained using test data are satisfactory and shows that the system can be used as an effective tool to classify children with disability and obtaining adequate information before consulting a specialist.

1 Introduction

Disability in children is one of the challenging issues that have increased the interest of the researchers in various directions. The symptoms observed at birth of a child along with mother's health condition during pregnancy seem to be very important factors to determine the types of disability in children. The presence of the same symptoms may result in different types of disability. In such cases, it is difficult to carry out proper treatment or impart special education for such children looking at the symptoms. The aim of this paper is to work on types of disability in children that depends on such combination of symptoms. In this paper, an attempt has been made to apply Artificial Neural Networks (ANN) for classification of disability in children based on the combination of symptoms. ANNs are nonparametric statistical tools designed using bio-inspired approaches [1]. ANNs have been configured for application in different fields as diverse as engineering technology to financial forecasting. The primary inspiration of the working principle of the ANN is created by mimicking the functioning of the human brain. ANNs can learn input patterns and use this knowledge to make predictions. Such an approach has been applied in this paper to classify the type of disability and relating them to the symptoms observed at birth of a child along with mother's health condition during pregnancy. The advantage of such a system is derived from the strengths of the ANN. ANNs are known to be adaptive, robust and can learn process data irrespective of methods of creation. All these spin-offs can be associated with this approach of predicting child disabilities using ANN. The paper proposes a Multi Layer Perceptron (MLP)-based ANN model for classifying types of disability among children. The model comprises of a single input layer which correspond to different symptoms at birth of a baby and maternal health during pregnancy and outputs corresponding to different types of disability. The preliminary results obtained using test data are satisfactory and shows that the system can be used as an effective tool to classify children with disability and obtaining adequate information before consulting a specialist.

The rest of the paper is organized as follows: Sect. 2 provides a background of the work. It covers types of disabilities, a brief review of certain important literature and an introduction to ANN. The proposed approach of disability classification using ANN is covered in Sect. 3. It includes a brief description of the studies area, data, methodology, system model and experimental details. Section 4 includes experimental results and related discussion. Section 5 concludes the description and provides a future direction.

2 Background

Disability is more a matter of perception than a barrier. It is much more difficult to measure disability in children than in adults. Furthermore the younger the child, the harder is to understand, articulate and measure disability. In fact disability is a very different concept for infants, toddlers, pre-school children, school-age children, and adolescents. Whereas play is the usual activity for the children aged 3–5 and attending school for those aged 5–17, the usual activity of infants and toddlers is unclear and hard to measure [2]. In medical terminology, disability is an impairment causing difficulty to carry out normal social roles. According to International Classification of Functioning, Disability and Health (ICF), 'disability' is 'an umbrella term for any or all of the components: impairment, activity limitation and participation restriction, as influenced by environmental factors' [3]. Briefly we can say that some of the babies are deficient physically and/or mentally from birth or due to some problems before and after birth. They are unable to actualize their potential for physical growth, intellectual behaviour and social development. These babies are grouped as disabled children or 'children with disability'. To confer them the sense of equality with the non-disabled ones, they are now-a-days also called 'differently abled'.

2.1 Types of Disability

Different types of disabilities, which may be due to genetic condition, an illness, environmental causes or an accident, may be classified as:

1. Mentally challenged or mental retardation (MR)
2. Cerebral Palsy (CP)
3. Communicative disorders (DSL)
4. Learning disability (LD)
5. Attention deficit hyperactive disorder (ADHD) and
6. Childhood autism or Autism

These disabilities are also termed as Developmental Disabilities [4], which are a group of disorders resulting from injury to the developing brain. A child with Intelligence Quotient below 70, significant limitations in two or more areas of adaptive behavior, and evidence that the limitations became apparent before the age of 18 is diagnosed as a mentally challenged child. CP is a group of non-progressive but often changing motor impairment syndromes secondary to lesions or anomalies of brain arising in early stages of its developments [4]. Communication disorder is a type of speech and language disorders which refers to problems in communication and in related areas such as oral motor function [5]. ADHD is characterized by poor ability to attend a task, motor over-activity and impulsivity. The boys to girl's ratio of ADHD children are 6:1 [4]. Autism is a brain development disorder characterized by impaired social interaction and communication, and by restricted and repetitive behavior. These signs all begin before a child is three years old [4]. Learning disability includes those children who get difficulty in coping with general academic skills. These are neurological differences in processing information that severely limit a person's ability to learn in a specific skill area.

2.2 Review

Morris [6] has reviewed the conceptual and operational limitations of classic approaches in classification of learning disability. He found that through the new development of more reliable and valid classification method will remove many of the present problems in clinical and research endeavors with learning disabled children.

Suresh and Raja [7] have applied Functional Magnetic Resonance Imaging (fMRI) through image processing techniques to classify SLD (specific learning disability), to determine depth of severity, degree of recovery and therapy.

Kohli and Prasad [8] have proposed a systematic approach for identification of dyslexia and to classify or analyze potential cases more accurately and easily by the use of ANN and found satisfactory results.

Jain et al. [9] have used single layer perceptron based artificial neural network model for diagnosing learning disability using curriculum based test conducted by special educators in medical environment. They found comparable experimental results on detection measures like accuracy, sensitivity and specificity.

Cohen et al. [10] have used ANN technology in classification of autism among children. They compare ANN method with simultaneous and stepwise linear discriminant analysis. They found neural network methodology is superior to discriminant function analysis both in its ability to classify groups (92 % vs. 85 %) and to generalize to new cases that were not part of the training sample. It has been observed that several researchers have applied ANN for classification of disability in children with encouraging results.

2.3 Artificial Neural Network (ANN)

ANN is an information processing paradigm that is inspired by the way biological nervous systems, such as the brain, process information. A mathematical analogy named the McCulloch-Pitts neuron (1943) [1] is the primary processing unit of the ANN. These artificial neurons or perceptions have the ability to learn patterns applied to them. Several such artificial neurons when are placed in multiple layers constitute the Multi Layer Perceptron (MLP) which is trained using (error) back propagation (BP) algorithm. Generally, the MLP may have an input, an output and a hidden layer for which it develops the ability of robust, model free, non-linear and adaptive processing. The output of a MLP with one hidden layer is given as:

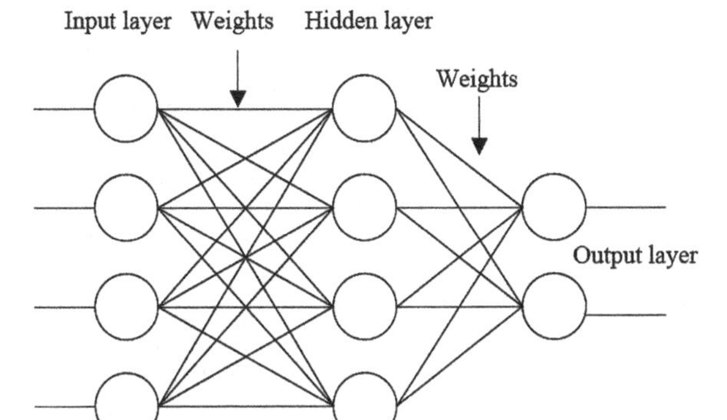

Fig. 1 Generic ANN

$$O_{jx} = \sum_{j=1}^{M} \sum_{i=1}^{N} f_{ji}\{[w]_{ji}.[x] + b_i\} \quad (1)$$

where f_{ji} represents the ith activation function in the jth layer, w_{ji} is the weight matrix for jth layer value between the ith hidden neuron and **x** input patterns with b_i bias. Figure 1 shows a generic ANN.

3 Proposed Approach of Classification of Disability in Children in Assam

Here, we briefly describe an approach of classifying disabilities in children of certain areas of the north-east (NE) Indian state of Assam. Specifically, we use an ANN based approach to identify disabilities in children using data collected from field study.

3.1 Classification of Children with Disabilities in the Studied Area

In Assam, there are over half a million persons with disabilities [11]. In comparison to this, the number of specialized institutions catering to these children is insufficient. Guwahati, the largest city in Assam and NE India, have a few institutions devoted to the care, education and treatment of children with disabilities, particularly with development disabilities. We have not found any reference to studies on classifications of the children with disabilities based on their symptoms in the studied area i.e. the geographical territory of Assam.

Table 1 Source of data

Sl. Num.	Name of source	Type of organization	Cases studied
1	National Institute of Public Cooperation and Child Development (NIPCCD), North Eastern Regional Centre, Guwahati	Institution under the Govt. of India	283
2	Mon Bikash Kendra, Guwahati, Assam, India	A special school run by an NGO—Guwahati Mental Welfare Society	124
3	Shishu Sarothi, Guwahati, Assam, India	A centre for rehabilitation and training for multiple disability, under the aegis of the Spastics Society of Assam, Assam, India	111

3.2 Data

Data on disabled children are collected from the records of the institutions mentioned in Table 1. The data included at-birth, pre- and post-natal condition of the baby and mother's health during pregnancy are considered.

3.3 Methodology

A total of 518 sets of data collected from the special schools/institutions are converted into a form suitable for the ANN to process. A coding scheme of the input sample data is developed using Boolean logic so that the patterns fed to the ANN are sequences of 1s and 0s. Seven critical symptoms identified as inputs for the problem. From the collected data, a coded input data set is formed depending upon the symptom observed. If a specific symptom is present 1 is assigned to the relevant column, while a 0 is placed if absent. This makes the input sample have values only of 1s and 0s. Two broad sets of pattern ensembles are formed. The first set is used for training the ANN. A few more data sets are formed with variation of $\pm 30\,\%$ compared to the training set. This set is called validation set used to check the level of training of the ANN. Another set is formed for testing the ANN which has $\pm 50\,\%$ variation compared to the training set. It is used to check the ability of the ANN extensively and make inferences regarding its ability to provide prediction with variations in input patterns. The complete set of 518 data is taken for classification of child disability. The input data set consists of the following seven symptoms at birth and maternal health during pregnancy:

- Mother's physical problem during pregnancy,
- Mother's mental stress during pregnancy,
- Birth cry,
- Birth weight,
- Baby's problem within a short period after birth,
- Psychomotor problem and
- Speech problem.

The target data consists of the following six outputs conforming to six types of disability:

- MR,
- CP,
- DSL,
- LD,
- ADHD and
- Childhood autism or autism.

A Multi-Layer Perceptron (MLP) with a single hidden layer is used in this study (Fig. 2), as it is known to be a strong function approximator for prediction and classification problems. The size of the input layer is seven as there are that many number of symptoms considered. The output layer includes six neurons as we have identified six types of disabilities. The size of the hidden layer is fixed with respect to number of neurons in the input layer and by following the considerations summarized in Table 2. It shows the performance obtained during training by varying the size of the hidden layer.

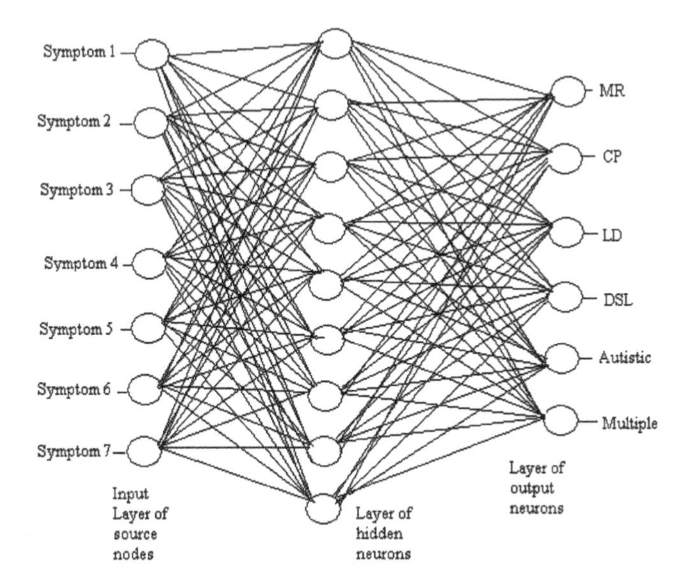

Fig. 2 MLP used in this study

Table 2 Performance variation after 1000 epochs during training of an ANN with variation of size of the hidden layer

Case	Size of hidden layer ($\times$ input layer)	MSE attained	Precision attained in %
1	0.75	1.2×10^{-3}	87.1
2	1.0	0.56×10^{-3}	87.8
3	1.25	0.8×10^{-4}	87.1
4	1.5	0.3×10^{-4}	90.1
5	1.75	0.6×10^{-4}	89.2
6	2	0.7×10^{-4}	89.8

The case where the size of the hidden layer taken to be 1.5 times to that of the input layer is found to be computationally efficient. hence, the size of the hidden layer is fixed at 1.5 times to that of the input layer.

The MLP is trained using BP algorithm to classify the available data into six classes with mean square error (MSE) convergence goal of 10^{-4}. The other performance criteria used is the classification and prediction accuracy expressed in percentage. The ANN is the critical element of the system and hence requires appropriate considerations for its configuration. Several trials are made before fixing the experimental configuration.

The ANN is formed by using the parameters as detailed in Table 3 and trained with different methods available in Matlab 7.5 viz. *Traingd*, *Traingdm*, *Traingda* and *Traingdx* to find out which training method will be the best for our data set. The method *Traingd()* is a network training function that updates the weight and biases used in the process according to gradient descent back propagation. The method *Traingdm()* is a ANN training function that updates the weight and biases according

Table 3 ANN configuration specifications

Case	Item	Description layer
1	Type	Multi layer perceptron
2	Data size	518 data of 7×6 size each
3	Input vector size	7
4	Class	6
5	Number of hidden layers	1
6	Learning rate	0.2 to 0.8
7	Activation function	*Logsig–Tansig–Logsig*
8	MSE convergence goal	10^{-4}
9	Training type	*Traingda*, Method of selection of ANN training type is as fixed from considerations shown in Table 4

to gradient descent with momentum back propagation. The method *Traingda()* is a network training function that updates the weight and biases used in the process according to gradient descent with adaptive learning rate back propagation. The method *Traingdx()* is another ANN training function that updates the weight and biases used in the process according to gradient descent with momentum and adaptive learning rate back propagation.

3.4 System Model

The entire work is summarized by a flow chart depicted in Fig. 3. The work consists of the following segments:

- Collection of data from field visits,
- Coding of the collected data for making it suitable for the ANN,
- Training the ANN and validating its performance,
- Testing the ANN and
- Recording response of ANN and enumeration of responses as per input.

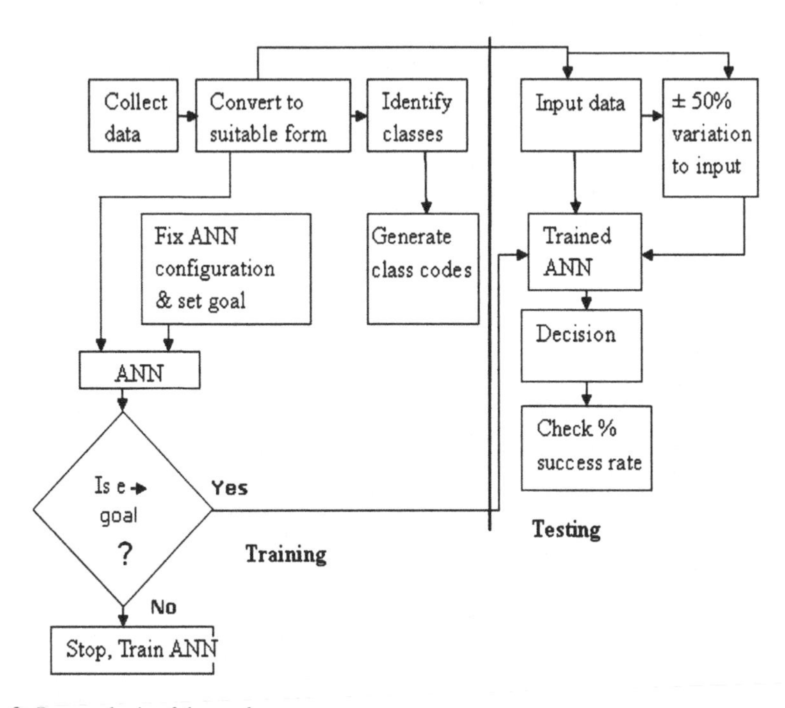

Fig. 3 Process logic of the work

4 Experiential Results and Discussion

The implementation of the system has two critical aspects—first training and next the testing. Application of ANN for classification of children with disability based on probable symptoms shows that such a tool is effective and reliable. Over 500 data sets are used to configure an ANN and design a non-parametric system which learns applied patterns. The ANN could handle input variations up to $\pm 50\,\%$ and provide satisfactory success rates in predicting classes of input data sets related to child disabilities. Experiments are performed as per the process logic summarized by Fig. 3. A range of experiments are performed initially to determine the most suitable training method for the ANN. It is necessary to maintain the efficiency of the system at the optimum level. The suitability of the training method is determined by the computational cost and the accuracy of classification. These two aspects are taken into consideration while fixing the training method of the ANN. A few number of training methods are adopted to see the effectiveness of the ANN. The values in Table 4 indicate that the *traingdx* is marginally faster than *traingda*, but the latter shows greater success-rates consistently while making class-wise discrimination. This is due to the fact that faster learning shown by *traingdx* at times though is desirable, but leads to false decision making as well. Hence, *traindga* is selected for training the ANN for carrying out the work. Table 3 gives the details of the parameters used for training the ANN with *traingda* method. If smaller learning rate is taken, ANN learns slowly but with a larger learning rate the training takes place at greater

Table 4 Selection of appropriate training method and training results

SL Num.	Epochs	MSE	Time (s)	Remark
traingd	300	0.28	6.93	Goal not reached
	500	0.37	3.45	
	1000	0.32	6.84	
	1500	0.27	9.98	
	2000	0.15	14	
	3000	0.22	20.2	
	4000	0.17	28.2	
	5000	0.2	33.89	
traingdm	300	0.44	2.34	
	500	0.27	3.7	
	1000	0.22	6.97	Goal not reached
	1500	0.18	10.34	
	2000	0.16	13.45	
	3000	0.15	20.95	
	4000	0.15	29	
	5000	0.16	33.9	
traingda	300	0.0008	2.3	Goal reached after 364 (av.) epochs and in 3 s
traingdx	300	0.0006	2.1	Goal reached after 322 (av.) epochs and in 2.57 s

Table 5 ANN training validation quantified with % of success in classification with variation in input samples between ±50 %

Variation	% of success	Variation	% of success
−0.5	50	0.05	100
−0.45	66.7	0.1	100
−0.35	77.8	0.15	100
−0.3	94.8	0.2	100
−0.25	100	0.25	100
−0.2	100	0.3	100
−0.15	100	0.35	100
−0.1	100	0.4	100
−0.05	100	0.45	100
0	100	0.5	100

pace. The slow learning helps in generalizations while a larger learning rate creates memorization. The latter case is undesirable [1]. The ANN thus trained is found to be useful in classifying the pattern of disability with a minimum error limit of 10^{-4}. This helps the ANN to become robust and reliable. It can deal with the classification problem effectively if relevant information recorded at child birth is included in the patterns applied to the system and mapped it to the appropriate class using the coding logic fed to the ANN.

The training results are summarized in Tables 4 and 5 and Fig. 4. The success rate at −50 % is lowest at around 50 % while beyond −30 % the success rate is nearly constant around 95–100 % for average reading of 10 runs per sample set. The rsults shown in Fig. 4 are generated by applying variations in the data upto ±50 %. At −50 % variation compared to the training data, the ANN shows a success rate of around 50 % for each of the ten trials when trained with the method *traingda*. This margin improves considerably and at around −30 %, the success rate of 94.8 % is achieved. The ANN's training, however, handles all subsequent variations comfortably and provides success rate of 95–98 % consistently. This reflects the ability of the system to handle input sample variations. It thus can be considered to be a robust, non-parametric prediction system.

To the best of our knowledge, no reported results, derived from equivalent or similar systems using known statistical techniques, are available for comparing the present outcome. The currently available statistical techniques are not used for similar pattern classification applications. Hence, the present work shall serve as a benchmark for subsequent developments in this direction.

5 Conclusion and Future Scope

In this paper, we discussed the details related to the application of an ANN based framework for classification of disability in children based on the combination of symptoms. The paper proposed a MLP-based model for classifying types of disability

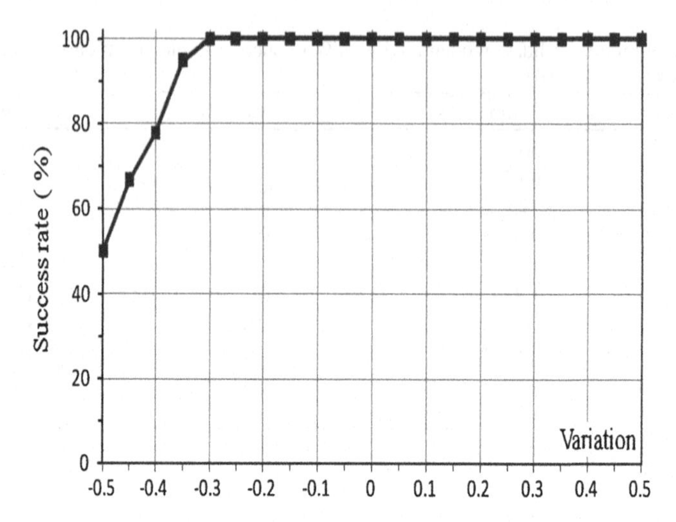

Fig. 4 Variation in data against success rate

among children. The model comprises of a single input layer which correspond to different symptoms at birth of a baby and maternal health during pregnancy and outputs corresponding to different types of disability. The preliminary results obtained using test data are found to be satisfactory and can be used as an effective tool to classify children with disability and obtaining adequate information before consulting a specialist.

Some probable factors of disability have been left out in this study for want of reliable data. These include family history, attempted abortion, use of prescription and non-prescription drugs during pregnancy, etc. Systematic studies can be conducted with collection of reliable primary data on the above. In some socio-environmental settings, consumption of alcohol and use of tobacco, even narcotic material may be not uncommon. If enough data on such factors can be gathered, the methods similar to that described here may be attempted. This study can also be extended to include the normal children to see the likelihood of a child acquiring disability based on the related factors.

References

1. Haykin, S.: Neural Networks, 2nd edn. Pearson Education, New Delhi (2006)
2. Eustis, N.N., Clark, R.F., Adler, M.C.: Research Agenda: Disability Data. Report of the U.S. Department of Health and Human Services (1995)
3. Disability World: UNICEF and Disabled Children and Youths. Disability World, No. 19. http://www.disabilityworld.org/06-08_03/children/unicef.shtml (2003)
4. Parthasarathy, A. (ed.): IAP Textbook of Pediatrics, 3rd edn. Jaypee Brothers, New York (2006)
5. Gleason, J.B., Ratner, N.B.: The Development of Language, 7th edn. Allyn and Bacon, Boston (2008)

6. Morris, R.D.: Classification of learning disabilities: old problems and new approaches. J. Consult. Clin. Psychol. **56**(6), 789–794 (1988)
7. Suresh, P., Raja, B.K.: A review on analysis and quantification of specific learning disability (SLD) with FMRI using image processing techniques. In: IJCA Proceedings on International Conference on VLSI, Communications and Instrumentation (ICVCI), vol. 5, pp. 24–29, Foundation of Computer Science (2011)
8. Kohli, M., Prasad, T.V.: Identifying dyslexic students by using artificial neural networks. In: Proceedings of the World Congress on Engineering, vol. 1, London, U.K. (2010)
9. Jain, K., Manghirmalani, P., Dongardivve, J., Abraham, S.: Computational diagnosis of learning disability. Int. J. Recent Trends Eng. **2**(3), 64–66 (2009)
10. Cohen, I.L., Sudhalter, V., Landong-Jimenez, D., Keogh, M.: A neural network approach to the classification of autism. J. Autism Dev. Disord. **23**(3), 443–466 (1993)
11. Census Report, Census of India, Govt. of India (2001)

[illegible]

[illegible] Intelligent [illegible]

[illegible] Vol. 24, pp. 1–16 [illegible]

[illegible]

Chapter 7
Evaluating Video and Facial Muscle Activity for a Better Assistive Technology: A Silent Speech Based HCI

Sridhar P. Arjunan, Wai C. Yau and Dinesh K. Kumar

Abstract There is an urgent need for having interfaces that directly employ the natural communication and manipulation skills of humans. Vision based systems that are suitable for identifying small actions and suitable for communication applications will allow the deployment for machine control by people with restricted limb movements, such as neuro-trauma patients. Because of the limited abilities of these people, it is also important that these systems have inbuilt intelligence and are suitable for learning about the user and reconfigure itself appropriately. Patients who have suffered neuro-trauma often have restricted body and limb movements. In such cases, hand, arms and the body movements may be impossible, thus head activity and face expression become important in designing Human computer interface (HCI) systems for machine control. Silent speech-based assistive technologies (AT) are important for users with difficulty to vocalize by providing the flexibility for the users to control computers without making a sound. This chapter evaluates the feasibility of using facial muscle activity signals and mouth video to identify speech commands, in the absence of voice signals. This chapter investigates the classification power of mouth videos in identifying English vowels and consonants. This research also examines the use of non invasive, facial surface Electromyogram (SEMG) to identify unvoiced English and German vowels based on the muscle activity and also provide a feedback to the visual system. The results suggest that video-based systems and facial muscle activity work reliably for simple speech-based commands for AT.

S. P. Arjunan (✉) · W. C. Yau · D. K. Kumar
Biosignals Lab, School of Electrical and Computer Engineering, RMIT University, GPO Box 2476, Melbourne, Victoria 3001, Australia
e-mail: sridhar.arjunan@rmit.edu.au

V. K. Mago and V. Dabbaghian (eds.), *Computational Models of Complex Systems*, Intelligent Systems Reference Library 53, DOI: 10.1007/978-3-319-01285-8_7,
© Springer International Publishing Switzerland 2014

1 Introduction

With the dynamic growth in communication technology, the demands for human computer interaction (HCI) techniques that enhance the flexibility for the users are increasing. New methods of computer control has focused on various types of body functions such as speech, emotions, bioelectrical activity, facial expressions, etc. The expression of emotions and speech play an important part in human interaction. Most of the facial movements result from either speech or the display of emotions; each of these has its own complexity [1].

Due to very dense information that can be coded in speech, speech based human computer interaction (HCI) can provide richness comparable to human-to-human interaction. Such systems utilize a natural ability of the human user, and therefore have the potential for making computer control effortless and natural. Speech-based systems are crucial for users with physical disabilities that hinder the use of hands [2]. Such systems are emerging as an attractive interface that provide the flexibility for users to control machines using speech. Speech recognition systems can be deployed in applications such as in-vehicle control systems [3], for better assistive technology (AT), security and surveillance systems, and telephony [4].

Human communication is multimodal and consists of a number of sensory information components. Based on this bio-inspired model, this study evaluates the feasibility of using non acoustic modalities to recognize speech in an effort to overcome the above mentioned limitations. Such silent speech recognition techniques require only the sensing of facial and speech articulators movement, without the need to sense the voice output of the speaker. There are a number of options available such as visual [5], recording of vocal cords movements [6], mechanical sensing of facial movement and movement of palate, recording of facial muscle activity [7], facial plethysmogram and measuring the intra-oral pressure.

The advantages of voiceless speech based system are such as (i) not affected by audio noise (ii) not affected by changes in acoustic conditions (iii) does not require the user to make a sound. The non acoustic cues contain far less classification power for speech compared to audio data and hence it is to be expected that such voiceless systems would have a small vocabulary of discrete commands.

This chapter evaluates the use of two non acoustic speech modalities—visual (video of lips and mouth) and facial muscle activity (surface electromyogram (SEMG)) for identification of silent speech. This study focusses on the classification of phonemes because phoneme-based system can be extended for word recognition by concatenating the phonemes to form words.

The main concern with such systems is the difficulty to work across people of different backgrounds and the main challenge is the ability of such a system to work for people of different native languages. This study compares the error in classification of the unvoiced English and German vowels by a group of German native speakers.The results of the previous studies [8, 9] suggest that it is difficult to recognize large vocabulary continuous speech using only video data. This chapter investigates the reliability of visual speech information in classifying a limited vocabulary of sub auditory speech consisting of English consonants.

The chapter is organized as follows: Sect. 2 discusses Video based speech analyser and Sect. 3 describes the facial muscle acivity based speech recognizition. Section 4 presents the methodology on using video data and facial muscle activity. Section 5 reports on the results and discussion. Section 6 presents a summary of this research work.

2 Video Based Speech Analyser

This research examines a vision-based technique to recognize English vowels and consonants based on the visible facial movement. A typical visual speech recognition(VSR) technique consists of three stages: (i) video processing for facial motion segmentation, (ii) visual feature extraction and (iii) classification. Based on the literature [10–12] visual speech features extracted from the videos can be categorized into shape-based, pixel-based and motion-based features. The shape-based features rely on the shape of the mouth.

This chapter investigates the classification accuracy of a novel VSR technique based on motion features extracted using spatial-temporal templates (STT) in English phoneme recognition. A visual speech model based on Moving Picture Experts Group 4 (MPEG-4) standard is used to map the consonants to the different patterns of facial movement. Figure 1 shows a block diagram of the proposed visual speech recognition technique.

2.1 Video Processing to Segment Facial Movement

A motion segmentation technique is used to extract the facial movements in the video data and represent the movement using a 2-D grayscale image (spatial-temporal

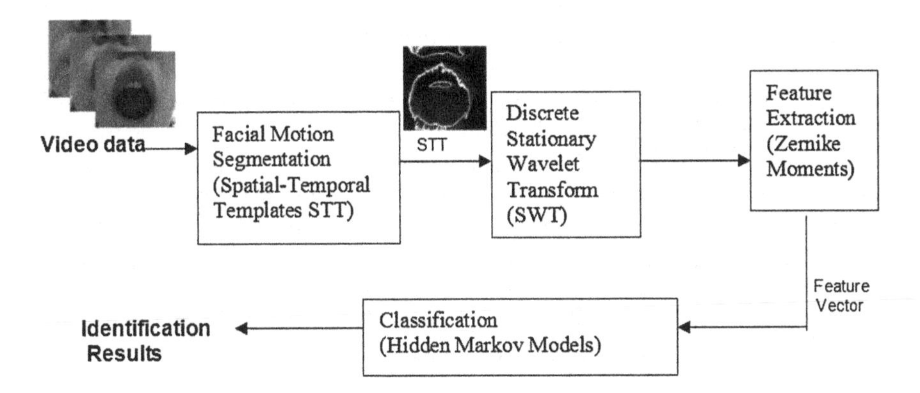

Fig. 1 Block diagram of the proposed visual speech recognition technique

templates(STT)) [5]. STT shows where and when facial movements occur in the image sequence [13].

STT are generated by using accumulative image difference approach. Image subtraction is applied on the video of the speaker by subtracting the intensity values between successive frames to generate the difference of frames (DOF). The DOFs are binarised using an optimum threshold value determined through experimentation. The delimiters for the start and stop of the motion are manually inserted into the image sequence. The intensity value of the STT at pixel location (x, y) of tth frame is defined by

$$STT_t(x, y) = max \bigcup_{t=1}^{N-1} B_t(x, y) \times t \tag{1}$$

where N is the total number of frames of the video. $B_t(x, y)$ represents the binarised version of the DOF of frame t. In Eq. 1, $B_t(x, y)$ is multiplied with a linear ramp of time to implicitly encode the temporal information of the facial motions into the STT. By computing the STT values for all the pixels coordinates (x, y) of the image sequence using Eq. 1 will produce a grayscale image (STT) where the brightness of the pixels indicates the recency of motion in the image sequence.

2.1.1 Issues Related to the Facial Movement Segmentation

STT is a view sensitive motion representation technique. Therefore the STT generated from the sequence of images is dependent on factors such as:

1. position of the speaker's mouth normal to the camera optical axis
2. orientation of the speaker's face with respect to the video camera
3. distance of the speaker's mouth from the camera (which changes the scale/size of the mouth in the video data)
4. small variation of the mouth movement of the speaker while uttering the same consonant

This technique proposes the use of discrete stationary wavelet transform (SWT) to obtain a transform representation of the STT that is insensitive to small variations of the mouth and lip movement. While the classical discrete wavelet transform (DWT) is suitable for this, DWT results in translation variance [14] where a small shift of the image in the space domain will yield very different wavelet coefficients. The translation sensitivity of DWT is caused by the aliasing effect that occurs due to the downsampling of the image along rows and columns [15]. SWT restores the translation invariance of the signal by omitting the downsampling process of DWT, and results in redundancies.

2-D SWT at level 1 is applied on the STT to produce a spatial-frequency representation of the STT. The 2-D SWT is implemented by applying 1-D SWT along the rows of the image followed by 1-D SWT along the columns of the image. SWT decomposition of the MHI generates four images, namely approximation (LL), horizontal detail coefficients (LH), vertical detail coefficients (HL) and diagonal detail coefficients (HH) through iterative filtering using low pass and high pass filters.

The approximate image is the smoothed version of the STT and carries the highest amount of information content among the four images. LH, HL and HH sub images show the fluctuations of the pixel intensity values in the horizontal, vertical and diagonal directions respectively. The image moments features are computed from the approximate sub image.The proposed technique adopts Zernike moments as the region-based features to represent the SWT approximate image of the STT.

2.2 Extraction of Visual Speech Features

Zernike moments are one of the feature descriptors used in classification of image patterns [16, 17]. Zernike moments have been demonstrated to outperform other image moments such as geometric moments, Legendre moments and complex moments in terms of sensitivity to image noise, information redundancy and capability for image representation [18]. The proposed technique uses Zernike moments as visual speech features to represent the SWT approximate image of the STT.

Zernike moments are computed by projecting the image function $f(x, y)$ onto the orthogonal Zernike polynomial V_{nl} of order n with repetition l is defined within a unit circle (i.e.: $x^2 + y^2 \leq 1$) as follows:

$$V_{nl}(\rho, \theta) = R_{nl}(\rho)e^{-\hat{j}l\theta}; \quad \hat{j} = \sqrt{-1} \tag{2}$$

where R_{nl} is the real-valued radial polynomial; $|l| \leq n$ and $(n - |l|)$ is even.

The main advantage of Zernike moments is the simple rotational property of the features [16]. The absolute value of Zernike moments is invariant to the rotational changes of mouth in the videos [5]. Zernike moments are also independent features due to the orthogonality of the Zernike polynomial Vnl [18]. For the Zernike moments to be orthogonal, the approximate image of the STT is scaled to be within a unit circle centered at the origin. The square approximate image of the STT is bounded by the unit circle . The center of the image is taken as the origin and the pixel coordinates are mapped to the range of the unit circle i.e.: $x^2 + y^2 \leq 1$. Figure 2 shows the square-to-circular transformation performed for the computation of the Zernike moments that transform the square image function $(f(x, y))$ in terms of the x-y axes to a circular image function $(f(\rho, \theta))$ in terms of the i-j axes.

The proposed approach uses the absolute value of the Zernike moments, $|Z'_{nl}|$ as the rotation invariant features of the SWT of STT. The absolute value of Zernike moments are rotation invariant [16]. By including higher order moments, more information of the STT can be represented by the Zernike moments features. However, this inherently increases the size of the features and makes it prone to noise. An optimum number of Zernike moments need to be selected to trade-off between the

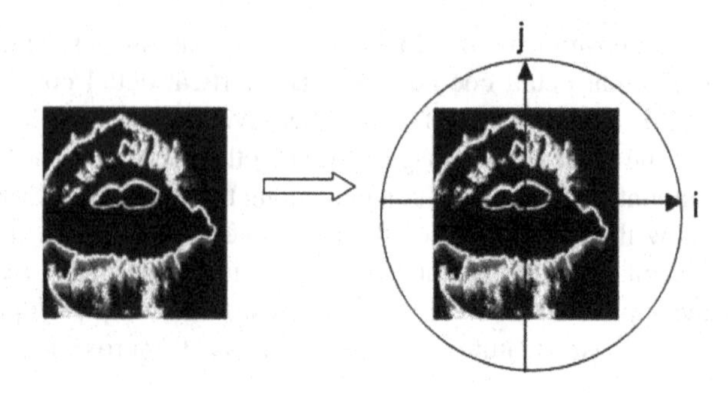

Fig. 2 The *square*-to-*circular* transformation of the SWT approximation of STT

dimensionality of the feature vectors and the amount of information represented by the features. 49 Zernike moments that comprise of 0th order moments up to 12th order moments have been used as features to represent the approximate image of the STT for each consonant [19].

3 Facial Muscle Activity Based Speech Recognition

The aim of this research is to analyse and classify the features of facial muscle activity recorded during utterance of vowels for giving simple commands to the computer. Such a system could be used to easily control the cursor on the small screen of a computer or mobile device by providing for the four directions and one click commands. For this study, the first step is to determine the role of the facial muscles in the production of speech.

3.1 Face Movement Related to Speech

The face can communicate a variety of information including subjective emotion, communitive intent, and cognitive appraisal. The facial musculature is a three dimensional assembly of small, pseudo-independently controlled muscular lips performing a variety of complex orfacial functions such as speech, mastication, swallowing and mediation of motion [20]. Both speech and emotions need a higher level specification of the controlling parameters of the facial muscles. The parameterization used in speech is usually in terms of phonemes. The required shape of the mouth and lips for the utterance of the phonemes is achieved by the controlled contraction of the facial muscles that is a result of the activity from the nervous system [21].

One difficulty with speech identification using facial movement and shape is the temporal variation when the user is speaking complex time varying sounds. With the intra and inter subject variation in the speed of speaking, and the length of each sound, it is difficult to determine a suitable window, and when the properties of the signal are time varying, this makes identifying suitable features for classification less robust.

The other difficulties also arise from the need for segmentation and the identification of the start and end of movement if the movement is complex. In such a system, using moving Root Mean Square (RMS) threshold, the temporal location of each activity can be identified. By having a stationary set of parameters defining the muscle activity for each spoken event, this also makes the system have very compact set of features, making it suitable for real time classification.

3.2 Features of SEMG

Surface Electromyogram (SEMG) is the non-invasive recording of muscle activity. It can be recorded from the surface using electrodes that are stuck to the skin and located close to the muscle to be studied. SEMG is a gross indicator of the muscle activity and is used to identify force of muscle contraction, associated movement and posture [22]. Using an SEMG based system, Chan et al. [23] demonstrated the presence of speech information in facial myoelectric signals. Kumar et al. [24] have demonstrated the use of SEMG to identify the unspoken sounds under controlled conditions. Recent studies have also reported the use of facial sEMG for various HCI applications [7, 25, 26].

While it is relatively simple to identify the start and the end of the muscle activity related to the vowel, the muscle activity at the start and the end may often be much larger than the activity during the section when the mouth cavity shape is being kept constant, corresponding to the vowel. The temporal location of the start and the end of the activity is identifiable using moving window RMS. The other issue is the variation in the inter-subject because of variation in the speed and style of utterance of the vowel. To overcome this issue, this research recommends the use of the integration of the RMS of SEMG from the start till the end of the utterance of the vowel.

4 Methods

4.1 Experiments Using Visual Video Data

Experiments were conducted to test the proposed visual speech recognition technique. Nine consonants that form the viseme model of English consonants according to the MPEG-4 standard are used in the experiment (Table 3).

4.2 Video Data Acquisition and Processing

Video data was recorded using an inexpensive web camera in a typical office environment. This was done towards having a practical voiceless communication system using low resolution video recordings. The reason that the experiments were conducted in an office environment (as opposed to noise-free studios) is to test the robustness of the visual system in one of the real life situations.

The camera focused on the mouth region of the speaker and was kept stationary throughout the experiment. The following factors were kept the same during the recording of the videos:

- window size of the camera
- view angle of the camera
- background
- illumination

180 video files with the resolution of 240×240 pixels were recorded and stored as true color (.AVI) files. The frame rate of the AVI files was 30 frames per second. One STT was generated from each AVI files. SWT at level-1 using Haar wavelet was applied on the STTs and the approximate image (LL) was used for analysis. 49 Zernike moments have been used as features (as explained in Sect. 2.2 to represent the SWT approximate image of the STT.

4.3 Experiments Using Facial Muscle Activity Signals

Experiments were conducted to test the performance of the proposed speech recognition from facial SEMG for two different languages, German and English. The experiments were approved by the Human Experiments Ethics Committee of the university. In controlled experiments, participants were asked to speak while their SEMG were recorded. SEMG recordings were visually observed, and the recordings with any artefact—typically due to loose electrodes or movement—were discarded. During these recordings, the participants spoke three selected English vowels (/a/, /e/, /u/) and three selected German vowels (/a/, /i/, /u/). Each vowel was spoken separately such that there was a clear start and end of its utterance. The experiment was repeated ten times for each language. A suitable resting time was given between each experiment. The participants were asked to vary their speaking speed and style to obtain a wide training set.

4.4 Facial EMG Recording and Preprocessing

The participants in the experiment were native speakers of German with English as their second language. Four channel facial SEMG was recorded using the recommended recording guidelines [27]. A four channel, portable, continuous recording

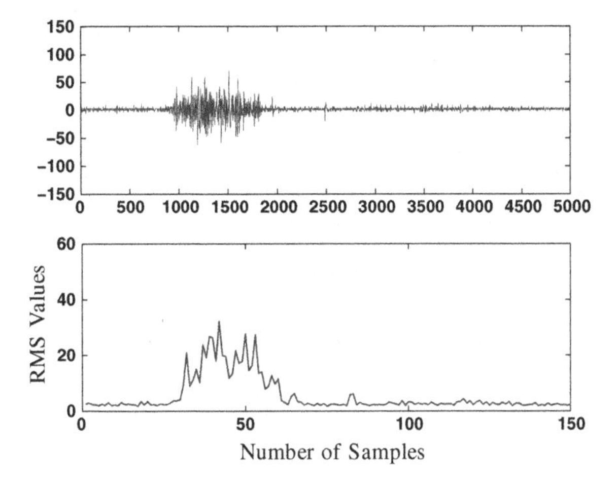

Fig. 3 Raw SEMG signal and its RMS plot

MEGAWIN instrument (Mega Electronics, Finland) was used for this purpose. Raw signal was recorded at a rate of 2000 samples/ second. Ag/AgCl electrodes (AMBU Blue sensors from MEDICOTEST, Denmark) were mounted on appropriate locations close to the selected facial muscles.

It is impractical to consider the entire facial muscles and record their electrical activity. In this study, only *four* facial muscles have been selected—*Zygomaticus Major*, *Depressor anguli oris*, *Masseter*, and *Mentalis* [27]. The recordings were visually observed, and all recordings with any artefact were discarded. Figure 3 shows the raw EMG signal recording and its RMS values plotted as a function of time represented by sample number.

4.5 Data Analysis

The first step in the analysis of the data was to identify the temporal location of the muscle activity. Moving root mean square (MRMS) of the recorded signal with a threshold of 1 sigma of the signal was applied for windowing and identifying the start and the end of the active period [28]. A Window size of 20 samples corresponding to 10 ms was used for computing the MRMS. The start and the end of the muscle activity were also confirmed visually.

MRMS values of SEMG between the start and the end of the muscle activity was integrated for each of the channels. This provided a four long vector corresponding to the overall activity of the four channels for each vowel utterance. This data was normalised by computing a ratio of integrated MRMS of each channel with respect to channel 1. This ratio is indicative for the relative strength of contraction of the

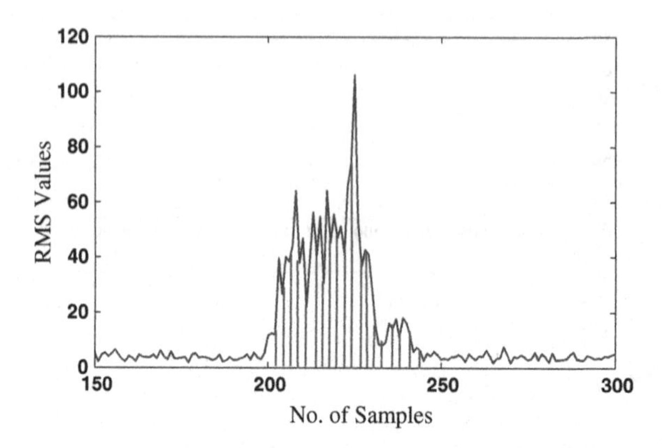

Fig. 4 An example of the computation of the integral of RMS of SEMG

different muscles and reduces the impact of inter-experimental variations. Figure 4 is an example of the computation of the integral of RMS of SEMG.

4.6 Classification of Visual and Facial EMG Features

For classification, supervised neural network approach was used with the parameterized data resulting in a vector for each utterance. The ANN consisted of two hidden layers with 20 nodes in both layers. Sigmoid function was used as the threshold decision. ANN was trained with gradient descent algorithm using a momentum with a learning rate of 0.05 to reduce the likelihood of local minima. Finally, the trained ANN was used to classify the test data. This entire process was repeated for each of the participants.

Random Sub sampling cross-validation method [29] was used to determine the mean classification accuracy of the normalized features of facial SEMG. The training and testing of different random sub samples using ANN was repeated for different times. The final classification accuracy is the average of individual estimates.

5 Results and Discussion

5.1 Facial Muscle Activity

Table 1 shows the ANN classification results on the test data using weight matrix generated during training for English vowels, and Table 2 lists these values for German vowels. These results indicate that the mean classification accuracy of the integral RMS values of the EMG signal yields better recognition rate of vowels

Table 1 Mean classification accuracies for English vowels

Vowels	Mean classification accuracy		
	Participant 1 (%)	Participant 2 (%)	Participant 3 (%)
/a/	73.33	83.33	80
/e/	76.67	76.67	83.33
/u/	100	100	100

Table 2 Mean classification accuracies for German vowels

Vowels	Mean classification accuracy		
	Participant 1 (%)	Participant 2 (%)	Participant 3 (%)
/a/	86.67	83.33	83.33
/i/	96.67	80	76.67
/u/	100	100	100

for three different participants, when it is trained individually. The results indicate that this technique can be used for the classification of vowels for the native and foreign language—in this case—English and German. This suggests that the facial muscle activity-based system is able to identify the differences between the styles of speaking of different people at different times for different languages.

5.1.1 Variation in Classification Error for Native and Foreign Language

Figures 5 and 6 show the variation in error rate for German and English vowels. The error rate in classification accuracy for a foreign language (English) is marginally higher compared with the native language (German). This is due to the muscle pattern remaining same during the utterance of the native language and changes during the utterance of the foreign language. The variation is high for German vowels /a/, /i/ and English vowels /a/, /e/ and there is no variation for the vowel /u/ in both German and English language.

The results indicate that the proposed method using activities of facial muscles for identifying silently spoken vowels is technically feasible from the view point of error in identification. The investigation reveals the suitability of the system for English and German, and this suggests that the system is feasible when used for people speaking their own native language as well as a foreign language.

The results also indicate that the system is not disturbed by the variation in the speed of utterance. The recognition accuracy is high, when it is trained and tested for a dedicate user. Hence, such a system could be used by any individual user as a reliable human computer interface (HCI). This method has only been tested for limited vowels.

The promising results obtained in the experiment indicate that facial muscles movement represents a suitable and reliable method for classifying vowels of single

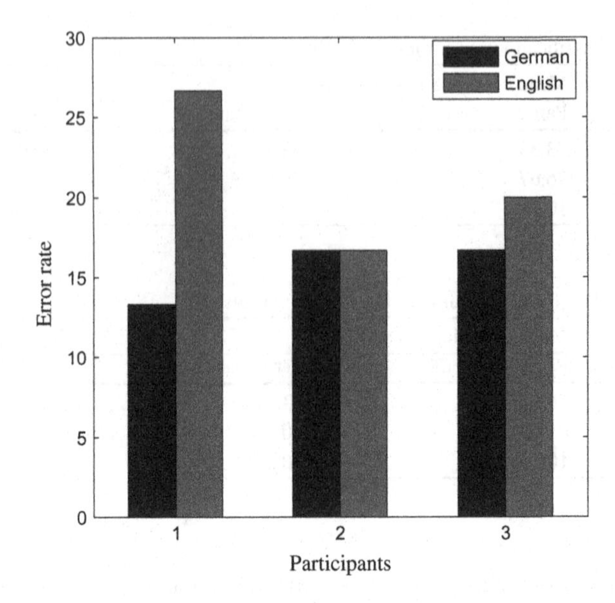

Fig. 5 Bar plot of the classification error rate for English and German Vowel-/a/

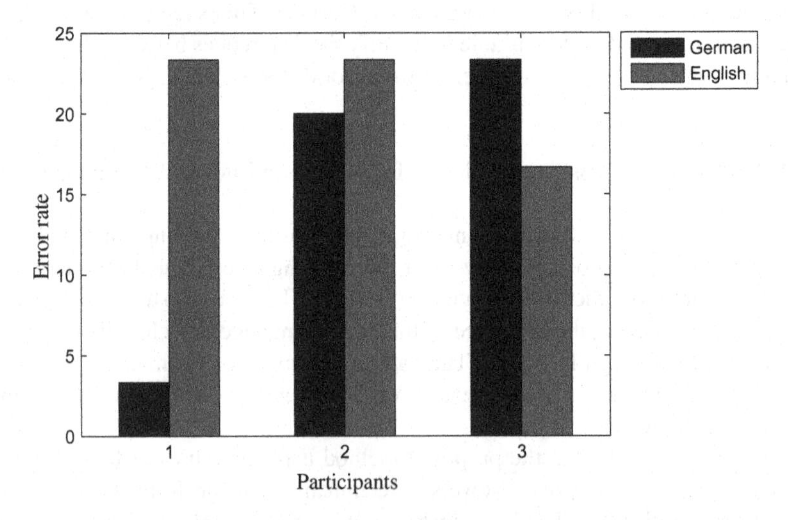

Fig. 6 Bar plot of the classification error rate for German vowel-/i/ and English Vowel-/e/

user without regard to speaking speed and style for different languages. It should be pointed out that the proposed technique based on facial muscles activity does not provide the flexibility of continuous speech, but for a limited dictionary of discrete phones, which is appropriate for simple voice control based AT systems. Furthermore, the results suggest that such a system is suitable and reliable for simple commands for human computer interface when it is trained for the user.

5.2 Visual Speech Recognizer

The classification accuracies of the HMM trained using visual features are tabulated in Table 3. The average recognition rate of the proposed vision-based system is 88.2 %. The results indicate that the proposed technique based on motion features is suitable for recognition English phonemes.

Based on the results, the proposed technique is highly accurate for vowels classification using the motion features. An average success rate of 97 % is achieved in recognizing vowels. The classification accuracies of consonants are slightly lower. Table 4 shows the differences in error rates for vowels and consonants. The results indicate that the consonants are less distinguishable than vowels using visual speech features.

The classification errors can be attributed to the inability of vision-based techniques to capture the occluded movements of speech articulators such as glottis, velum and tongue. For example, the tongue movement in the mouth cavity is either partially or completely not visible (occluded by the teeth) in the video data during the pronunciation of alveolar and dental sounds such as /t/, /T/ and /n/. The STTs of /t/, /T/ and /n/ do not contain the information of the occluded tongue movements. This is a possible reason for the higher error rates of 20, 30 and 60 % for these three consonants as compare to the average error rate of 12 % for all visemes. Consonantal sounds with similar facial movements may cause ambiguities that affect the performance of visual speech recognizer.

Table 3 Mean recognition rates of the visual speech recognizer based on viseme model of MPEG-4 standard

Viseme	Recognition rate (%)
m	95
v	90
T	70
t	80
g	85
tS	95
s	95
n	40
r	100
A:	100
e	100
I	95
Q	95
U	95

Table 4 Mean recognition rates for English vowels and consonants

Vowels/consonants	Recognition rate (%)
Vowels	97
Consonants	83.33

To compare the results of the proposed approach with other related work is inappropriate due to the different video corpus and recognition tasks used. In a similar visual-only speech recognition task (based on the the 14 visemes of MPEG-4 standard) reported in [30], a similar error rate was obtained using shape-based features (geometric measures of the lip) extracted from static images. Nevertheless, the errors made in our proposed visual system using motion features are different compare to the errors reported in [30]. This indicates that complementary information exist in static and dynamic features of visual speech.

For example, our proposed system has a much lower error rate in identifying visemes /m/, /t/ and /r/ by using the facial movement features as compare to the results in [30]. This shows that motion features are better in representing phones which involve distinct facial movements (such as the bilabial movements of /m/). The static features of [30] yield better results in classifying visemes with ambiguous or occluded motion of the speech articulators such as /n/. The results demonstrate that a computationally inexpensive system which can easily be developed on a DSP chip for silent speech based assistive technology.

6 Conclusion

This chapter reported the feasibility of sub auditory speech recognition approaches based on video data and measuring of the facial muscle contraction using non-invasive surface Electromyogram (SEMG). Application of this includes removal of any disambiguations caused by the acoustic noise for Human computer based assistive devices.

The presented investigation focused on the identification of English and German vowels using SEMG and the recognition accuracy for the SEMG-based system is high, when it is trained and tested for a dedicate user, in both German and English. This study also examined a vision-based technique to recognize English vowels and consonants. The experimental results of the visual approach demonstrate that the visual speech information can be used to reliably classify a set of English phonemes. One basic application for such a system is for disabled user with speech impairment to give simple commands to a machine, which would be a helpful and typical application of assistive technology (AT).

References

1. Ursula, H., Pierre, P.: Facial reactions to emotional facial expressions: affect or cognition? Cogn. Emot. **12**(4), 509–531 (1998)
2. Feng, J., Sears, A., Karat, C.: A longitudinal evaluation of hands-free speech-based navigation during dictation. Int. J. Hum. Comput. Stud. **64**, 553–569 (2006)

3. Kuhn, T., Jameel, A., Stuempfle, M., Haddadi, A.: Hybrid in-car speech recognition for mobile multimedia applications. In: IEEE Vehicular Technology Conference, Houston, TX., USA, 2009–2013 (1999)
4. Starkie, B., 2001. Programming Spoken Dialogs Using Grammatical Inference. AI 2001: Advances in Artificial Intelligence: 14th International Joint Conference on Artificial Intelligence. Adelaide, Australia
5. Yau, W.C., Kumar, D.K., Arjunan, S.P. 2006. Visual Speech Recognition Method Using Translation. Scale and Rotation Invariant Features, IEEE International Conference on Advanced Video and Signal based Surveillance, Sydney, Australia
6. Dikshit, P. S., Schubert, R. W., 1995. Electroglottograph as an additional source of information in isolated word recognition. Fourteenth Southern Biomedical, Engineering Conference, 1–4.
7. Arjunan, S., Kumar, D.K., Weghorn, H., Naik, G.: Facial muscle activity patterns for recognition of utterances in native and foreign language: testing for its reliability and flexibility. In: Mago, V., Bhatia, N. (eds.) Cross-Disciplinary Applications of Artificial Intelligence and Pattern Recognition: Advancing Technologies, pp. 212–231. Information Science Reference, Hershey (2012)
8. Potamianos, G., Neti, C., Gravier, G., Senior, A. W.: Recent Advances in Automatic Recognition of Audio-Visual Speech. Proceedings of IEEE, vol. 91 (2003)
9. Hazen, T.J.: Visual model structures and synchrony constraints for audio-visual speech recognition. IEEE Trans. Audio Speech Lang. Process. **14**(3), 1082–1089 (2006)
10. Petajan, E.D.: Automatic Lip-reading to Enhance Speech Recognition. IEEE Global Telecommunication Conference (1984)
11. Kaynak, M.N., Qi, Z., Cheok, A.D., Sengupta, K., Chung, K.C.: Audio-visual modeling for bimodal speech recognition. IEEE Trans. Syst. Man Cybern. B Cybern. **34**, 564–570 (2001)
12. Adjoudani, A., Benoit, C., Levine, E.P.: On the integration of auditory and visual parameters in an HMM-based ASR, Models, Systems, and Applications, Speechreading by Humans and Machines, pp. 461–472. (1996)
13. Bobick, A.F., Davis, J.W.: The recognition of human movement using temporal templates. IEEE Trans. Pattern Anal. Mach. Intell. **23**, 257–267 (2001)
14. Mallat, S.: A Wavelet Tour of Signal Processing. Academic Press (1998)
15. Simoncelli, E.P., Freeman, W.T., Adelson, E.H., Heeger, D.J.: Shiftable multiscale transform. IEEE Trans. Inf. Theory **38**, 587–607 (1992)
16. Khontazad, A., Hong, Y.H.: Invariant image recognition by zernike moments. IEEE Trans. Pattern Anal. Mach. Intell. **12**, 489–497 (1990)
17. Teague, M.R.: Image analysis via the general theory of moments. J. Opt. Soc. Am. **70**, 920–930 (1980)
18. Teh, C.H., Chin, R.T.: On image analysis by the methods of moments. IEEE Trans. Pattern Anal. Mach. Intell. **10**, 496–513 (1988)
19. Yau, W.C., Kumar, D.K., Weghorn, H.: Motion Features for Visual Speech Recognition. In: Liew, A., Wang, S. (eds.) Visual speech recognition: Lip segmentation and mapping, pp. 388–415. Medical Information Science Reference, Hershey (2009)
20. Lapatki, G., Stegeman, D.F., Jonas, I.E.: A surface EMG electrode for the simultaneous observation of multiple facial muscles. J. Neurosci. Methods **123**, 117–128 (2003)
21. Parsons, T. W.: Voice and speech processing, 1st edn, McGraw-Hill Book Company, New York (1986)
22. Basmajian, J.V., Deluca, C.J.: Muscles alive: Their functions revealed by electromyography. 5th edn. (1985)
23. Chan, D.C., Englehart, K., Hudgins, B., Lovely, D.F.: A multi-expert speech recognition system using acoustic and myoelectric signals. 24th Annual IEEE EMBS/BMES Conference (2002)
24. Kumar, S., Kumar, D.K., Alemu, M., Burry, M.: EMG based voice recognition. In: Prooceddings of intelligent sensors, sensor networks and information processing conference (2004)
25. Arjunan, S.P., Weghorn, H., Kumar, D.K., Naik, G., Yau, W.C.: Recognition of human voice utterances from facial surface EMG without using audio signals. Enterp. Info. Syst. Lect. Notes Bus. Info. Process. **12**(6), 366–378 (2009)

26. Tuisku, O., Surakka, V., Vanhala, T., Rantanen, V., Lekkala, J.: Wireless Face Interface: Using voluntary gaze direction and facial muscle activations for human computer interaction. Interact. Comput. **24**(1), 1–9 (2012)
27. Fridlund, A.J., Cacioppo, J.T.: Guidelines for human electromyographic research. J. Biol. Psychol. **23**(5), 567–589 (1986)
28. Freedman, D., Pisani, R., Purves, R.: Statistics. Norton College Books, New York (1997)
29. Gutierrez-Osuna, R., Lecture 13: Validation. http://research.cs.tamu.edu/prism/lectures/iss/iss_l13.pdf (Last Access: June 2012)
30. Foo, S.W., Dong, L.: Recognition of visual speech elements using hidden Markov models. Lect. Notes Comput. Sci. Springer-Verlag **2532**, 607–614 (2002)

Chapter 8
Modeling Returned Biomedical Devices in a Lean Manufacturing Environment

Camille Jaggernauth

Abstract With the projected rise in the senior population, the use of biomedical devices play an indispensable role in the monitoring of the elderly, for staving off the onset of complications as well as providing peace of mind for family members and relief for care-givers. This chapter examines the modeling of the social aspect of biomedical device manufacturing in a lean manufacturing environment. The social aspect includes customer satisfaction with the product as it relates to increased sales. The type of modeling used is fuzzy cognitive.

1 Introduction

The chapter discusses modeling repair, replace criteria in a lean manufacturing environment for biomedical devices. The motivation for this work is to produce a fuzzy cognitive map (FCM) which would indicate areas of improvement in the manufacturing process as well as predictors of demand for the biomedical devices industry where the bottom line is a social aspect of complex system modeling i.e. an increase in customer satisfaction causing an increase in sales. The customer identified here could either be an elderly person or care-giver purchasing the device for an elderly person.

No previous work has been done on modeling reverse logistics for biomedical devices in a lean manufacturing environment with respect to the use of these devices for elder care with focus on customer satisfaction.

Figure 1 below shows the system FCM. The system map comprises FCM submaps: role of biomedical devices in health-care A, sales of biomedical devices B, biomedical device lean manufacturing C, reverse logistics in lean manufacturing D, repair or replace criteria for biomedical devices E and customer service F.

C. Jaggernauth (✉)
SFU, 8888 University Drive, Burnaby, B.C.V5A 1S6, Canada
e-mail: jaggerna@sfu.ca

V. K. Mago and V. Dabbaghian (eds.), *Computational Models of Complex Systems*, Intelligent Systems Reference Library 53, DOI: 10.1007/978-3-319-01285-8_8,

© Springer International Publishing Switzerland 2014

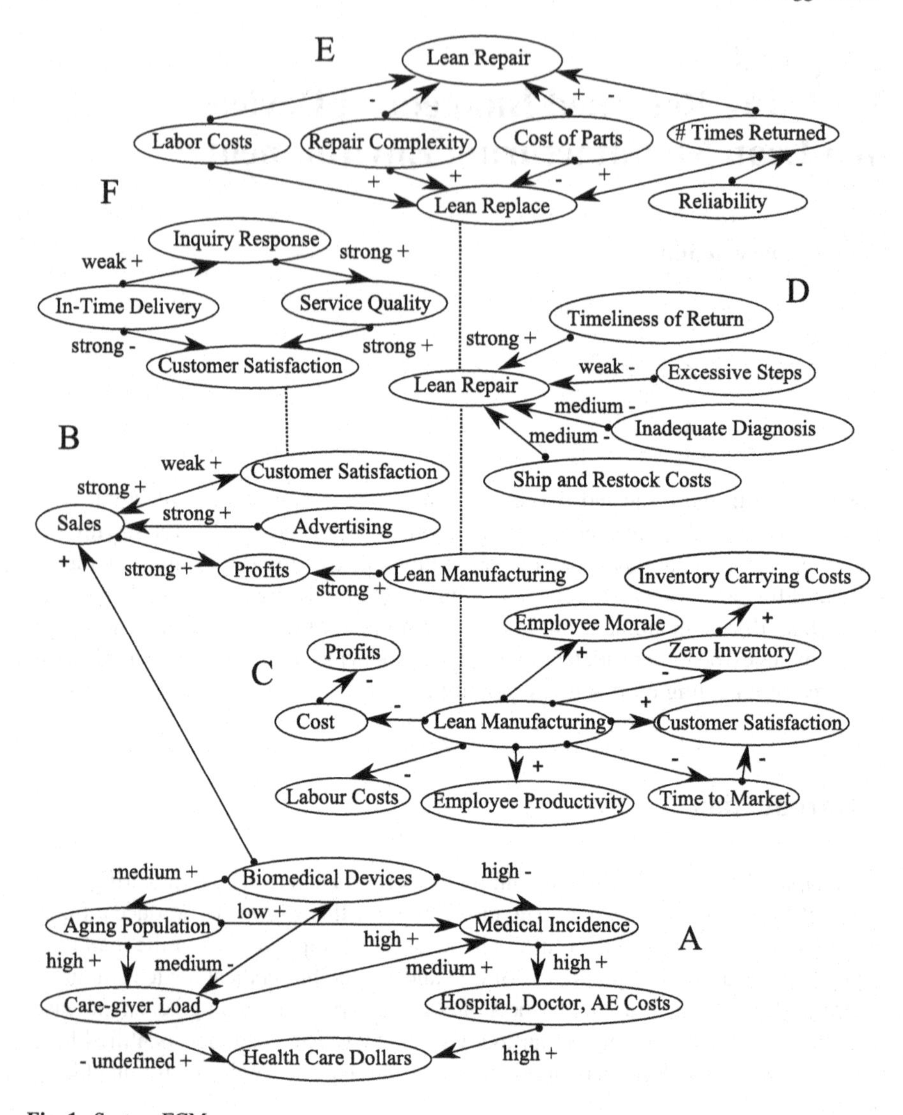

Fig. 1 System FCM

Between the FCM sub-maps, shared concepts or concepts that are subsets of each other are connected with dashed lines, regular concepts were connected using causal relationships.

Some of the FCM sub-maps can be alternatively represented as hierarchical decision trees (repair or replace criteria for biomedical devices) or information process flow maps (lean manufacturing, reverse logistics in lean manufacturing).

Despite the above mentioned potential for alternate representations, fuzzy cognitive mapping was used to ensure uniform analysis of the complex composite system, to enable the use of fuzzy linguistic weights and feed-back functionality, and to provide the ground-work where future work could include observing the dynamic performance of the system if real value weights are substituted.

Additional future work for this simple, non-exhaustive FCM could include the role of the biomedical devices in human activity modeling and modeling user satisfaction with the device.

Fuzzy cognitive maps rely on literature review to provide the weight of the edges or consultation with experts (including case-studies and consideration of relevant practical experience) [14, 18]. Both sources of information were utilized when deriving the FCM sub-maps. The system FCM is a visual representation for all stake-holders, including project analysts, planners and management to see the process flow and how different decisions [14] (e.g. resource allocation, investment into research for better quality products, providing loaners while products are being repaired) will affect the desired outcome i.e. increased customer satisfaction resulting in increased sales.

2 Fuzzy Cognitive Maps

2.1 Background

Fuzzy cognitive maps have a background in fuzzy logic and neural networks. FCMs represent causality (relative change in a concept causes a relative change in another concept) but are not a logic approach to causality. The fuzzy part allows for degrees of causality and the neural networks part allows for the presentation of one or multiple inputs at an artifical neuron (concept) for processing into an output value. FCMs are represented diagramatically as a signed directed graph with feedback [4]. Augmented FCMs have been built from fuzzified decision trees [19].

FCMs are an effective tool for modeling complex social systems with sufficient interacting parameters and ill-structured domains e.g. decision making. FCMs have even been developed for applied sciences and engineering applications. FCMs facilitate ease of use and low time requirement [14, 18]. Because of the above mentioned characteristics, FCM was chosen to model the chapter's research topic.

2.2 Weights, Causality and Analysis

The type of fuzzy cognitive map modeling used is described in *The use of fuzzy cognitive maps to simulate the information systems strategic planning systems* by Kardaras et al. [14]. The Kardaras model overcomes the problem of measuring real-numbered weights by using fuzzy linguistic weights and keeps the complexity in the lower levels while not sacrificing the clarity of concepts.

There are four weights $undefined < weak < moderate < strong$.

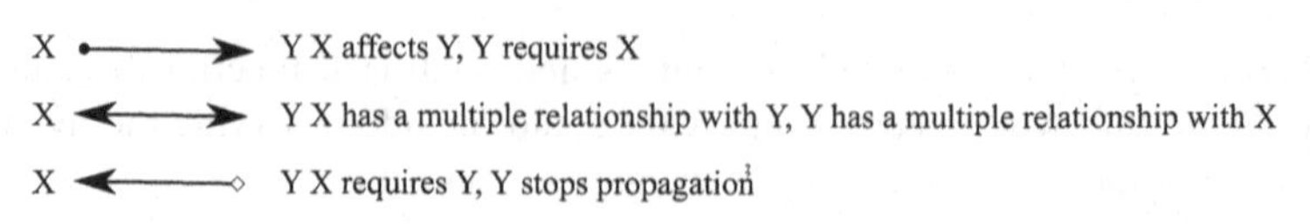

Fig. 2 Causality

There are four causal types *affects*, *requires*, *multiplies* and *stops*. If *X affects Y* then an increase/decrease in X results in an increase/decrease in Y. If *X requires Y* then an increase/decrease in X does not result in an increase/decrease in Y AND an increase/decrease in Y is required for an increase/decrease in X. If *X multiples Y* then an increase/decrease in X results in an increase/decrease in Y AND an increase/decrease in X requires an increase/decrease in Y. If *X stop Y* then an increase/decrease in X has no effect on Y (Fig. 2).

In the Kardaras model, decision analysis occurs by firstly identifying the causal paths I, followed by determining the polarity of the path S, the degree of belief ϕ and the most believed effect Δ.

The polarity of the path S is (+) if the number of negative polarity relationships is even or zero.

The degree of belief ϕ of the path is determined by minimum of the fuzzy linguistic weights along the path.

The most believed effect Δ is path which yields the maximum degree of belief. Detailed decision analysis is presented in Sect. 4.

2.3 Research Investigation

The chapter development proceeds by firstly developing the submaps of the system FCM (Fig. 1) in Sects. 3.1–3.5, and then presenting detailed simulations on the system FCM, including model validation in Sect. 4 followed by conclusions and future work in the discussion in Sect. 5.

3 Deriving the System FCM

This section develops the FCM sub-maps of the system FCM: role of biomedical devices in health-care A in Sect. 3.1, sales of biomedical devices B in Sect. 3.2, biomedical device lean manufacturing C and reverse logistics in lean manufacturing D in Sect. 3.3, repair or replace criteria for biomedical devices E in Sect. 3.4 and customer service F in Sect. 3.5.

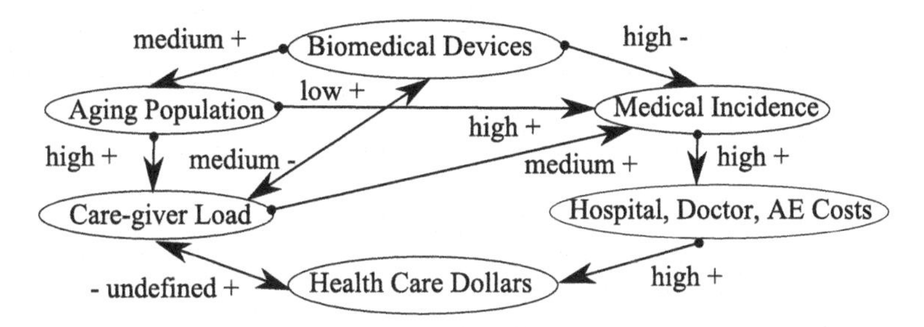

Fig. 3 Role of biomedical devices FCM A with fuzzy linguistic weights

3.1 Modeling the Role of Biomedical Devices in Health-care

This section details the development of the core FCM sub-map, the role of biomedical devices in health-care FCM (Fig. 3).

In the IEEE Spectrum article *The Doctor will see you ALWAYS* by Smith [21] the usage of biomedical devices is explored for ambient monitoring and specifically its role in chronic disease management (which eats up 75 % health care spending or US $1.9 trillion annually). The article highlights specific existing and potential future device usage scenarios. Highlighted in the article are a personal sleep coach (for monitoring sleep cycles, online if required) insulin pumps with glucose monitoring, wireless enabled scale, smart phone applications for chronic disease management and tele-health solutions (physiological, mental and medication monitoring) [11, 12, 17]. A 2008 trial run of a tele-health solution with veterans who experience multiple chronic conditions has found that device usage catches complications early on and has reduced hospital admissions by 19 % and days spent in health care facilities by 25 %.

Most recent population projections suggest that the senior population of British Columbia (B.C.), Canada, will double in the next 20 years. This will lead to increased usage of health care facilities, home and community care services (HCC—Home Care Community Services and Accommodation Environments) as well as social services [8, 20].

Elderly people are not only afflicted by physical problems associated with aging but also with mental illness (suicide, dementia, distress, stress) and substance abuse problems which require the use of health care systems for proper care [6].

Existing commercial solutions help family members monitor aging relatives as well as empowering professional care-givers by using sensors to track daily activity levels as well as falls. Daily as well as routine changes and detection of asymptomatic behavior could indicate medication complications, congestive heart failure worsening or depression [21].

Friends and family provide $2.7 billion in free care and are the primary health-care system in Canada [22–24]. The Statistics Canada 2008 report *Eldercare* estimates that 1 in 5 Canadians over 45 were caring for a senior in 2007 where 25 % of senior care-givers are seniors themselves. A 1999 US study has found elderly care-givers of spouses have a 63 % higher mortality and if a care-giver gets sick, more people enter the facility. Studies show stress of care-giving can have debilitating and harmful physical, mental and emotional consequences. In Canada, there is a lack of measures available to ease the financial, emotional and psycho-social repercussions on care-givers [16].

Based on the literature review done above and the information provided by experts the below preliminary fuzzy cognitive map with polarities was developed (Fig. 3). Because actual numeric information was available from the data sources, linguistic fuzzy cognitive weights were tentatively added to allow for further preliminary simulation analysis. As mentioned in the previous section, there is currently no governmental policies to ease the financial, emotional and psycho-social effect on care-givers, so this fuzzy linguistic weight is considered undefined.

3.2 Modeling the Sales of Biomedical Devices

Because the same concepts are present in both industries for the scope of this analysis, the FCM for sales of biomedical devices (Fig. 4) is a subset, simplified and inserted from the FCM for the car industry described in [25]. The FCM was updated to use fuzzy linguistic weights which were assigned based on existing numeric values and assuming a linear mapping. ($0 < weak <= 0.33, 0.33 < medium <= 0.67, 0.67 < strong <= 1.00$).

3.3 Modeling Lean Manufacturing

The following paragraphs detail the development of the lean manufacturing FCM for biomedical devices (Fig. 5) based on the key principles of lean manufacturing.

The Lean Manufacturing concept (or JIT Just in Time) was developed in mid 1970s by Toyota. Lean manufacturing techniques increase profitability by reducing costs according to how customers define value. Costs that do not add value are reduced or eliminated. Wastes pinpointed in lean manufacturing include transport, inventory, motion, waiting, overproduction, over processing and defects [1, 9]. Lean manufacturing leads to significant improvements in other areas including employee morale and productivity, customer satisfaction and faster time to market [2, 5, 15].

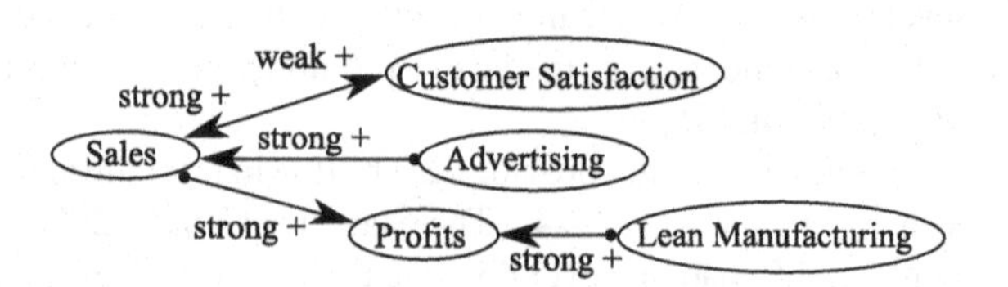

Fig. 4 Sales FCM B with polarities and fuzzy linguistic weights

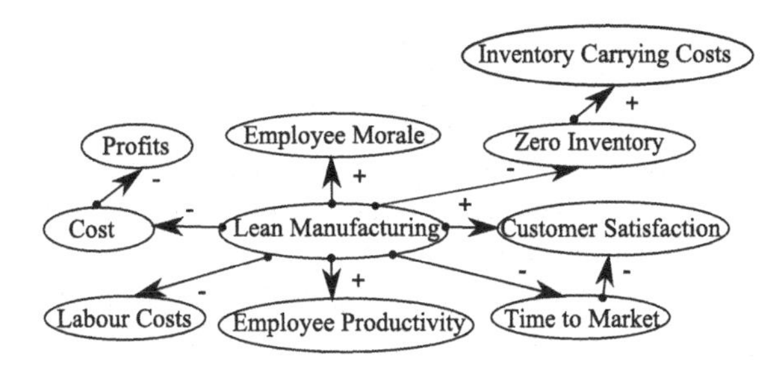

Fig. 5 Lean manufacturing FCM C with polarities

In Regular Production larger quantities are ordered or produced resulting in an average larger inventory (increased inventory-carrying charges) or smaller quantities are ordered (increased ordering costs). Comparatively in Lean Production the ideal production is quantity of one. While a lot size of one is not always feasible it can be used as a goal to focus on rapid adjustments and flexibility [2, 9, 15]. There are two types of lean production—replacement and spontaneous build-to-order. In replacement lean, the production parts are pre-built and pulled into assembly from kanban bins (kanban scheduling system what, when, how much to produce). Spontaneous build-to-order lean utilizes kanban and parts built on-demand from standard raw materials [1]. Based on the literature review done above and the information provided by consultant experts (see references), the below preliminary fuzzy cognitive map with polarities was developed (Fig. 5).

The following paragraphs detail the development of the lean repair FCM for biomedical devices (Fig. 6). Lean repair is a subset of lean manufacturing.

Researching Lean Repair yielded a case-study by a consulting company in reverse logistics (the return of a product from the user to the manufacturer or distributor because of a number of reasons including service contract stipulations, bad quality, and unmet customer expectations etc.) commissioned by a biomedical company and presented at a lean manufacturing conference [3]. Based on the case-study findings, the below lean goals in repairs were specifically targeted for the biomedical company:

- Ensuring timeliness of returned parts—increasing customer accountability on unreturned parts including regular customer notification (5th, 15th notification) with updates to management (30th, 45th day notification) as well as suggested

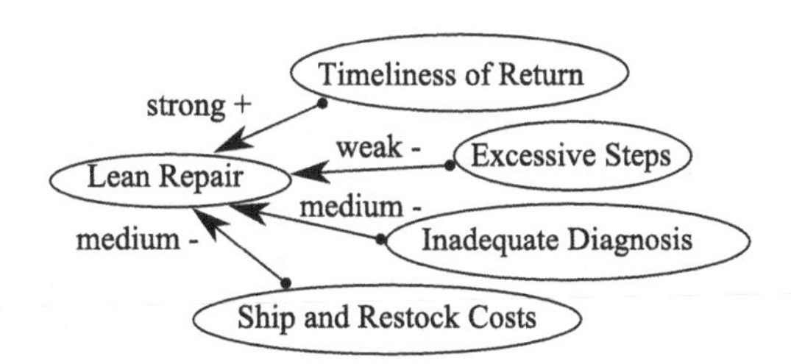

Fig. 6 Reverse logistics FCM D with polarities and fuzzy linguistic weights

hard deadlines for customer invoicing of parts, closing returns or RGA (returned goods authorizations) where there was no customer contact.

- Reducing excessive steps in returns process—more clear and concise verbiage in the voice recording messages, prompting the customer to have the appropriate RGA #s available in order to expedite the call and effective call redirection based on the customer input.
- Addressing inadequate root cause information—better documentation improved deficiencies of poor visibility and root cause analysis. Better training was provided on soliciting and writing root cause as well as improving the consistency and accuracy in the trouble-shooting process.
- Addressing excessive shipping and restocking costs—reducing the number of FedEx account numbers. Next day return replaced with second-day return. Restocking fee updated from a flat 20 % re-stocking fee to: 20 % for parts greater than $500 list price, $105 on items less than $500 list price.

Based on the experts viewpoint presented above including the actual numeric savings provided, the below preliminary fuzzy cognitive map with polarities was developed. Because actual numeric information was available from the experts, linguistic fuzzy cognitive weights were tentatively added to allow for further preliminary simulation analysis (Fig. 6).

3.4 Modeling Replace or Repair

In order to determine the repair or replace fuzzy cognitive map (Fig. 7), manufacturing and repair engineers from local industries as well as resident academic experts were consulted. The feedback indicated that typical repairs for electronics included field service wear and tear—broken, lifted electronic traces, components, firmware updates, poor quality—malfunctioning parts, poor diagnosis and marketing reasons.

In certain scenarios the cost of parts for electronic devices is inexpensive when compared to the cost of labor (a time-expensive and intensive operation would be hand-soldering fragile electronic traces in flexible microelectronics).

Additionally, depending on the complexity of the repair, replacing through rebuilding is cheaper than repair. If the device has been returned for repair multiple times,

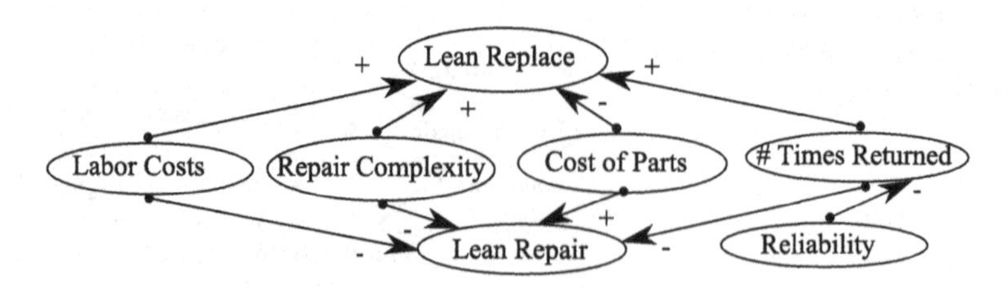

Fig. 7 Lean repair or replace FCM E with polarities

Fig. 8 Customer service FCM F with polarities and fuzzy linguistic weights

the best practice would be to replace the device, so as not to compound the perception of device unreliability and increase customer dissatisfaction. Figure 7 represents a polarities based fuzzy cognitive map (as opposed to a fuzzy linguistic weight based FCM).

3.5 Modeling Customer Service

The FCM for modeling customer service in a biomedical industry (Fig. 8) is a subset and is simplified and inserted from the FCM for customer service in [14] because the same concepts (product service in-time delivery, response to customer enquiry for product, customer service quality) are present in both industries for the scope of this analysis. The FCM was augmented to include the customer satisfaction concept by consulting with customer service and repair engineers from local industries.

4 Simulation: Forward and Backward

4.1 Forward Simulation

In the Kardaras model [14], forward simulation involves estimating the effects of change (increase or decrease) to a variable. The causal path only considers *affects* and *multiple* relationships. An example in this model would be to observe the effect of change on an increase in allocation of health care dollars. The concepts/variables are given below. Detailed decision analysis proceeds as per Sect. 2 and Table 1.

Following the *affects/multiple* causal path shows the reducing of the load on care-givers, resulting in reducing in the incidence of medical conditions in the care-givers and load on the health-care system. An additional causal path would be the reducing of load in the care-giver resulting in the reducing of the dependence on biomedical devices. A biomedical company observing a change in health care funding to support care-givers would have to invest in more product development and research in order to maintain sales and overall customer satisfaction with the product.

Table 1 Forward Simulation—increase in health-care dollars allocation

C1 = Health care dollars
C2 = Care-giver load
C3 = Medical incidence
C4 = Hospital, doctor, AE cost
C5 = Biomedical devices
C6 = Sales
C7 = Customer satisfaction
I_1 = (C1, C2, C3, C4)
I_2 = (C1, C2, C5, C6, C7)

4.2 Backward Simulation

In the Kardaras model [14], backward simulation is used to identify the actions required to achieve certain goals. Alternative paths must be defined by *requires* or *multiple* relationships. This section performs a backward simulation to increase the customer satisfaction variable on the system FCM (Fig. 1). The concepts/variables are given below. Detailed decision analysis proceeds as per Sect. 2, Tables 2 and 3.

For simulation purposes C7 = C5 as lean repair logistics is considered a subset of lean repair which is a subset of lean production/manufacturing, and the relationship between C7 and C1 is assumed to be strong because lean production/manufacturing based on the literature review is considered to be customer based or eliminating wastage based on customer perception of value [1, 9].

In the below case (Table 3) negative polarity S indicates an inverse relationship exists. The most believed effects Δ are the paths of strong degree of belief ϕ i.e. I_1, I_4.

It is observed that these paths are associated with timeliness of returned parts. From a planning perspective if resources need to be allocated or money invested to increase customer satisfaction it would best be done along these paths. From a repair versus replace analysis for focus on increasing customer satisfaction, timeliness is

Table 2 Concepts for backward simulation—increase in customer satisfaction

C1 = Customer satisfaction
C2 = Customer service quality
C3 = Response to customer enquiry for product
C4 = Product, service in-delivery time
C5 = Lean manufacturing
C6 = Time to market
C7 = Lean repair
C8 = Timeliness of returned parts
C9 = Excessive steps in return process
C10 = Inadequate root cause information
C11 = Excessive shipping and restocking

Table 3 Backward simulation analysis—increase in customer satisfaction

$I_1 = (C1, C4)$	$= \{\text{Strong} -\}$
$I_2 = (C1, C2, C3, C4)$	$= \{\text{Strong} +, \text{strong} +, \text{weak} +\}$
$I_3 = (C1, C6, C5)$	$= \{-, -\}$
$I_4 = (C1, C7, C8)$	$= \{\text{Strong} +, \text{strong} +\}$
$I_5 = (C1, C7, C9)$	$= \{\text{Strong} +, \text{weak} -\}$
$I_6 = (C1, C7, C10)$	$= \{\text{Strong} +, \text{medium} -\}$
$I_7 = (C1, C7, C11)$	$= \{\text{Strong} +, \text{medium} -\}$
$S^1 = -$	$\phi^1 = \text{Strong}$
$S^2 = +$	$\phi^2 = \text{Weak}$
$S^3 = +$	$\phi^3 = \text{Undefined}$
$S^4 = +$	$\phi^4 = \text{Strong}$
$S^5 = -$	$\phi^5 = \text{Weak}$
$S^6 = -$	$\phi^6 = \text{Medium}$
$S^7 = -$	$\phi^7 = \text{Medium}$

important, hence lean replacement or lean repair should be chosen based on which would allow the product to be delivered first and within the deadlines set in Sect. 3.3 and with consideration to the concepts listed in the repair versus replace FCM.

4.3 Practical Application

The B.C. (British Colombia, Canada) provincial budget 2012 [7] includes a tax credit for seniors for home renovations (to reduce the cost of structural changes) and a zero-credit GST/HST tax relief for medical devices prescribed by medical practitioners. A practical application of this model would be to predict the outcome on selected concepts (biomedical devices sales) based on these budgetary changes. The concepts/variables are given below. Detailed decision analysis proceeds as per Sect. 2 and Table 4.

Using *forward simulation*, it can be seen that two budgetary changes would result in an increase in the sales of biomedical devices through the increase of advertising (budget publication). The tax credit on renovations allows for more disposable income to spend on biomedical devices and the zero-credit GST/HST tax relief on medical devices would also provide greater incentive to purchase biomedical devices.

Table 4 Forward simulation analysis—effects of budgetary changes

C1 = Advertising
C2 = Sales
$I_1 = (C1, C2)$

Table 5 Prelimnary model validation

$I_{1survey}$ = (C1, C4)	= {Strong}
$I_{2survey}$ = (C1, C2, C3, C4)	= {Strong, strong, strong}
$I_{4survey}$ = (C1, C7, C8)	= {Strong, medium}
$I_{5survey}$ = (C1, C7, C9)	= {Strong, medium}
$I_{6survey}$ = (C1, C7, C10)	= {Strong, medium-strong}
$I_{7survey}$ = (C1, C7, C11)	= {Strong, medium-weak}
ϕ^1_{survey} = Strong	ϕ^1 = Strong
ϕ^2_{survey} = Strong	ϕ^2 = Weak
ϕ^4_{survey} = Medium	ϕ^4 = Strong
ϕ^5_{survey} = Medium	ϕ^5 = Weak
ϕ^6_{survey} = Medium strong	ϕ^6 = Medium
ϕ^7_{survey} = Medium weak	ϕ^7 = Medium

4.4 Validation

One way to validate the model (Table 5) would be to perform detailed surveys with customers, management and technical supervisors of biomedical companies to rate different concepts in the FCM and to compare the actual findings with model predictions. Concepts could be rated for importance and satisfaction and an adequate sample size could be used [10]. Additional in-company performance metrics could also be measured including cycle time, customer service, quality and costs and compared against model predictions [15].

Preliminary investigations into validating the model, involved designing a survey [13] to validate the strength of the links of the customer service and lean manufacturing FCM, as well as to query for additional concepts. A sample size of 25 was used and responses were compared against results of the backward simulation in Sect. 4. Participants had backgrounds in customer service, biomedical devices and lean manufacturing in customer, engineering and technical management roles.

Detailed decision analysis proceeds as per Sect. 2 and Table 5 above. Analysis of the validation results reveals that ϕ^1_{survey} is identical to ϕ^1. Additionally ϕ^6_{survey}, ϕ^7_{survey} are close to ϕ^6 and ϕ^7. Consequently Δ_{survey} is I_1, I_2 and I_6 compared to Δ I_1 and I_4. The survey introduces a new Δ_{survey} I_2 and I_6 which only marginally edges out I_4, I_5, I_7 (tied) for most believed effect. The conclusion could be drawn that better granularity could be achieved if the sample size was increased. Additionally preliminary validation involved only two sub-maps. Subsequent work could include the validation of the remaining sub-maps.

5 Discussion

In conclusion, a fuzzy cognitive map was developed to model replace/repair criteria for returned biomedical devices in a lean manufacturing environment. Fuzzy cognitive maps provided a framework which allowed planners to consider the complex interactions between business and manufacturing and perform analysis in a comprehensive and systematic manner. Fuzzy cognitive maps modeled the beliefs which stake-holders share with respect to the model's relationship. Forward and backward simulation were used to determine the potential of improving sales and customer satisfaction. It was found by forward simulation of the model that an increase in care-giver funding by the government would provide less of a burden on the health-care system while reducing the demand for biomedical devices. It was also found by backward simulation using this model that timeliness was key to customer satisfaction with repair experience of the device. Preliminary attempts at model validation showed correlation but a larger sample size would improve granularity.

Future work could include elaboration on the concepts, (variables) - here simple fuzzy cognitive maps were used. Real values for weights and transfer functions could be used instead of linguistic fuzzy weights to allow for the finding of an equilibrium state and to observe the dynamic behavior of the model. Further research into improving customer satisfaction could involve defining the customer the next person in the process (not only the end device user) [15]. Additional future work could include the role of biomedical devices in human activity modeling.

References

1. Anderson, D.: On demand lean production. In: Built-to-Order Consulting. http://www.halfcostproducts.com/lean.htm. Cited 23 Oct 2011
2. Bergmiller, G., et al.: Parallel models for lean and green operations. In: Industrial Engineering Research Conference (2009)
3. Boston Industrial Consulting: Case study in reverse logistics. In: IIE Lean Manufacturing Conference (2004)
4. Carvalho, J.: On the Semantics and the Use of Fuzzy Cognitive Maps in Social Sciences. IEEE World Congress on Computational Intelligence — WCCI. 2456–2461 (2010).
5. Dudley, A.: The application of lean manufacturing principles in a high mix low volume environment. MBA Thesis, MIT, Massachusetts (2005)
6. Government of British Colombia: Seniors mental health and substance use issues. In: Government of British Colombia publication. http://www.heretohelp.bc.ca/sites/default/files/images/seniors_mh.pdf. Cited 24 Oct 2011
7. Government of British Colombia: Fiscal discipline for a stable economy. In: Government of British Colombia (Canada) Budget 2012. http://www.bcbudget.gov.bc.ca/2012/highlights/2012_Highlights.pdf
8. Hare, W., et al.: A deterministic model of home and community care client counts in British Columbia. Health Care Manage. Sci. **12**(1), 80–99 (2009)
9. Inman, A.: Lean manufacturing and just-in-time production. In: Reference for Business, Encyclopedia for Business. http://www.referenceforbusiness.com/management. Cited 23 Oct 2011
10. Iqbal, T., et al.: Analysis of factors affecting the customer satisfaction level of the public sector in developing countries: an empirical study of automotive repair service quality in Pakistan. In: Global Conference on Innovations in Management, pp. 11–29 (2011)
11. Jaggernauth, C., et al.: Optimized, practical firmware design for a novel flexible wireless multi-sensor platform for body activity and vitals monitoring. In: IEEE International Conference on Consumer Electronics(2011)
12. Jaggernauth, C., et al.: Test firmware architecture for a flexible wireless physiological multi-sensor, SMC (2011)

13. Jaggernauth, C.: Modeling replace/repair criteria for returned biomdedical devices in lean manufacturing. http://www.surveymonkey.com/s/3MJSWQM
14. Kardaras, D., Karakostas, B.: The use of fuzzy cognitive maps to simulate the information systems strategic planning process. Inf. Softw. Technol. **41**, 197–210 (1999)
15. Machado, V., et al.: Modeling lean performance. In: IEEE ICMIT (2008)
16. OConnor, E.: Hardest job on the planet. The Province. 21 October 2011
17. Pantelopoulos, A., et al.: A survey on wearable sensor-based systems for health monitoring and prognosis. IEEE Trans. Syst. Man Cybern. Part C Appl. Rev. **40**, 1–12 (2010)
18. Papageorgiou, E., et al.: A fuzzy cognitive map based tool for prediction of infectious diseases. In: FUZZ-IEEE (2009)
19. Papageorgiou, E.: A novel approach on designing augmented fuzzy cognitive maps using fuzzified decision trees. FSICA (2009)
20. Schrier, D.: Consequences of an aging population, can existing levels of social services be sustained? In: Government of British Colombia Publication. http://www.bcstats.gov.bc.ca/data/pop/pop/agingpop.pdf. Cited 24 Oct 2011
21. Smith, J.: The doctor will see you always. IEEE Spectr. Mag. **48**(10), 56–62 (2011)
22. Stajduhar, K., et al.: Interviewing family caregivers: Implications of the caregiving context for the research interview. Qualitative Health Research (2012)
23. Stajduhar, K., et al.: Planning for end-of-life care: findings from the Canadian study of health and aging. Can. J. Aging **27**, 11–21 (2008)
24. Stajduhar, K., et al.: Situated/being situated: client and coworker roles of family caregivers in hospice palliative care. Soc. Sci. Med. **67**, 1789–1797 (2008)
25. Tsadiras, A.: Comparing the inference capabilities of binary, trivalent and sigmoid fuzzy cognitive maps. Inf. Sci. **178**, 3880–3894 (2008)

Chapter 9
EARLI: A Complex Systems Approach for Modeling Land-use Change and Settlement Growth in Early Agricultural Societies

Sonja Aagesen and Suzana Dragićević

Abstract Understanding the driving forces of historical land-use change can provide insights about the pressures experienced in present-day landscapes. This study develops a model to examine the spatio-temporal land-use changes and population responses of early agricultural communities under a variety of environmental and cultural conditions. Complex systems theory and geographical information systems (GIS) are integrated into the design of the model. The resulting Early Agricultural Resources and Land-use Investigation (EARLI) model couples agent-based modeling (ABM) and cellular automata (CA) techniques within a GIS framework. The model examines how both cultural and environmental factors influence land use change under multiple scenarios. Results from the simulations provide new insights that range from analysis of the interaction of human behavior within the local environment to the archaeological record.

1 Introduction

Land-use and land-cover change have been systematically studied for several decades [2, 12, 52]. These studies have focused primarily on present-day land-use problems and have developed common tools for understanding urban and environmental phenomena [21, 48, 61]. However, the analysis of ancient land-use change processes that span many centuries is poorly articulated in the literature for a number of

S. Aagesen
Spatial Interface Research Lab, Simon Fraser University, 8888 University Drive, Burnaby, BCV5A 1S6, Canada
e-mail: saagesen@sfu.ca

S. Dragićević (✉)
Spatial Analysis and Modeling Lab, Simon Fraser University, 8888 University Drive, Burnaby, BCV5A 1S6, Canada
e-mail: suzanad@sfu.ca

V. K. Mago and V. Dabbaghian (eds.), *Computational Models of Complex Systems*, Intelligent Systems Reference Library 53, DOI: 10.1007/978-3-319-01285-8_9,
© Springer International Publishing Switzerland 2014

reasons [28, 50]. On a practical level, there is limited continuous empirical data for long historical time depths to cover the extent of land use change once domestication of crops began. On a conceptual level, identifying the driving variables associated with human-environment interactions is a complex process, spatially explicit but with some degree of inherent uncertainty associated with the idiosyncratic nature of socio-cultural aspects of anthropogenic change [1].

The introduction of agriculture into early communities resulted in more sedentary lifestyles as people developed the local landscape to accommodate a new way of life. Turning virgin land into arable cropland was a dynamic process requiring numerous culturally localized decisions throughout. When put into action, these land-use decisions based on yearly interactions amongst weather, community needs, and land-use changes created complex spatial patterns that were intractable and unique for a given time period. Therefore the objective of this study is to develop the Early Agricultural Resources and Land-use Investigation (EARLI) simulation model in order to examine the spatio-temporal land-use changes and population responses of early agricultural communities under a variety of environmental and cultural conditions. The model's design integrates multi-criteria evaluation (MCE), geographical information systems (GIS) and complex systems theory, more particularly cellular automata (CA) and agent based modeling (ABM) approaches.

The development of the EARLI model combines GIS data from archaeological studies on early farming [10, 58, 64] as well as ethno-archaeological and ethnological investigations from a variety of climates and geographical locations [19, 37, 39, 63]. Ethnography provides a greater understanding of human-human and human-environment interactions [9, 29]. In studying human-environmental systems, non-unique social phenomena can be replicated like natural processes [7], but dynamic social events are complex and require many parameters [41].

Settlement remnants are usually all that remain from past agricultural communities. Community populations can be estimated from the settlement size using linear mathematical models [51], but the extent to which ancient farmers changed the surrounding landscape is difficult to approximate. The EARLI model uses local demographics and land-use change to simulate the process of early agricultural community growth, their maintenance, and possible decline. EARLI was developed to support the identification of factors that influence land-use changes in specific regions including climatic and environmental factors, and internal factors based on culture which represents "adjustments to a particular environment" [59]. The EARLI model also present exploratory tool for analysis of different cultural strategies and the influence of early societal controls like environment exploitation and birth-rate.

This study begins with a background of complex systems modeling and then outlines the method used to develop and implement the EARLI model. Simulation results are presented for both humid and arid climatic conditions. Sensitivity analysis of model parameters is then performed to determine reliability and aid in interpreting the results.

2 Theoretical Background

2.1 Agricultural Modeling in Archaeology

The adoption of agriculture is an important moment in prehistory and has been the subject of archaeological research for decades because it corresponds to a change in lifeways for a culture leading to societal complexification [23]. Typically, models of early agricultural strategies are built from three variables: population size, caloric needs, and agricultural yields [24, 58]. Expanding on this, Smith [58], incorporates parameters like reliance on agriculture, caloric intake, and calories per kilogram, but uses a rigidly defined agricultural zone to discover the maximum potential under a number of scenarios. Whereas Gregg [31], developed a model to consider how the annual harvest would be determined for the number of people/households once their needs were subtracted from the harvest. The problem with these approaches is that they are entirely mathematical and do not consider spatial or social components.

Since these models are not spatial or dynamic and did not provide temporal feedback, they were limited to outcomes based on input parameter derivatives. Furthermore, because all the contributing factors cannot be incorporated, most of the suggested and implemented models cannot depict land-use change or growth nor can they represent population dynamics. They are in essence compartmentalised models.

Recently, three less compartmentalized models have been developed encompassing both spatial and temporal data linked to anthropogenic responses. Davies [20] has produced a conceptual land-use model that takes into account possible feedback from both environmental and social factors over a long time depth. Louwagie's et al. [45] model focuses on land suitability within Rapa Nui (Easter Island), and sufficiently equates possible cultivated land extents with potential populations. Zhang et al. [66] consider both spatial and social aspects in terms of a catchment's natural environment and inter-site proximity, and uses principle component analysis (PCA) for understanding these correlations within arable agriculture in China. These models are still not dynamic and limit themselves to derived outcomes, rather than allowing for emergent patterns.

2.2 GIS and Multi-Criteria Evaluation

In many cases, Geographical Information Systems (GIS) can be used to include the spatial component of ancient agricultural field extents through simple regression programming based on settlement area. A common problem with GIS though, is its inability to deal adequately with the temporal components of environmental processes [26]. In this research, GIS was used to integrate the raster images of the landscape spatially. Additionally, GIS can be combined with Multi-Criteria Evaluation (MCE) to produce layers of information related to spatial suitability [14, 46].

MCE is an optimization approach that combines information from several criteria to create an index of evaluation. This is then applied to aggregated criteria layers to produce a suitability map. The criteria can be constraints or factors, with the weighting of each criterion considered in the final output. The MCE approach can be linked to a CA model decision-making [65]. In this study, GIS-based MCE is used to provide necessary information such as relevant to the CA model input and are the foundation of EARLI model.

2.3 Complexity Theory and Land-use Modeling

Complex systems theory considers a system to comprise many inter-related components such that the behavior of each component depends on the behavior of the others [11, 47]. The complex systems approach is one that features a large number of interacting components, like agents, whose aggregate activity is nonlinear and exhibits hierarchical self-organization under selective pressures [57]. The primary challenge in complexity theory is concerned with how basic processes can produce emergent patterns [2]. Emergent patterns can come from dynamic or complex system models of interacting component parts that do not represent a steady state [7]. This approach can model social dynamic phenomena that capture human-decision complexity, thus revealing cultural processes within their interactions [17, 47].

In modeling land-use patterns of early societies, nonlinear interactions do exist as the exchange between environments, human decisions, and geographic landscape provides both complexity and dynamism. Complex system modeling also permits the integration of synthetic and natural landscapes to examine an aspect of a real-world situation in a controlled setting [22].

2.4 Cellular Automata and GIS

Cellular automata are a dynamic class of complex system models in which a set of deterministic or probabilistic rules generate patterns across space within a system that is updated at discreet time intervals [5, 36]. This modeling technique is a bottom-up approach that can capture complex patterns better than GIS's overlay and map algebra functions [2]. CA operates at the local level and through progressive iterations it produces global patterns [56]. These cells independently vary their state based on the influence of the adjacent cells (neighborhoods) and a set of transition rules representing the logic of the process being modeled [62]. CA models can handle spatial and temporal dimensions; they are commonly used to model complex systems such as land-use change because of [4, 5, 21]. However, although in modeling human decision systems the transition rules should include stochastic elements [61] traditional CA is limited in its inclusion of social processes [38] and cannot capture feedback smoothly [52] or more in depth human-environment interactions. To date, CA has not been undertaken to explore dynamics within the studies of early agriculture.

2.5 Agent Based Models

Agent-based modeling is an inductive process that uses individual agents defined by behavioral rules that react to their environment resulting in emergent change over time [11]. ABM also offers a method of applying rules of interaction within systems helping to move beyond one-way cause and effect. The advantage of this is that behaviors are represented in a more natural way giving variation over both time and space, incorporating non-linear dynamics [2]. When coupled with GIS, ABM better represents both spatial patterns and temporal processes [11]. When combined with CA, its spatial simulation offers a promising degree of flexibility and provides a range of outcomes [52]. In land-use change studies, neither environmental nor agent-level factors are the single causality, but rather the interaction of the two [33]. The use of ABM method within the proposed EARLI model provides for the inclusion of social interactions through decision making.

Several agent-based models have been developed for understanding complex adaptive systems based on agricultural data and community needs. These include the Artificial Anasazi Project [22] in the American Southwest studying the Pueblo period (A.D. 800–1300) and the Patrimonial Household Model for ancient Egypt [43].

3 Methodology

The design of the EARLI model is divided into five parts: suitability map, model parameters, model stochastic parameters, model variables and transition rules. The suitability map based on MCE provides underlying information when determining land-use change. The model parameters provide input for the real-world data; the stochastic parameters provide values that simulate changeable conditions and community decisions. Model variables are calculated per time step using parametric data values at the previous time step ($t-1$) and with stochastic events. The agent-based decisions are made at the community level in response to population changes versus agricultural output, especially during times of environmental stress. The CA transition rules determine the land-use change potential based on suitability.

3.1 Suitability Map

The GIS–based, MCE-derived suitability map provides additional spatial information for the model to use as a component of the CA transition rules. Land suitability is based on factors such as soil-type, slope, land cover, availability of water, and distance to settlement. The constraints, where no agriculture is possible, are the rivers and the ridge tops. For clarity, the simulation uses uncomplicated factor layers and a simplified paleo-landscape as a suitability map. However, for case-study use the creation of the suitability map should be consistent with local environmental conditions. The value for the suitability of a cell is $f(Env_{suit})$ scaled between 0 and 1 derived from the equation based on Eastman et al. [27].

$$Env_{suit} = w_i x_i \quad (1)$$

where:
w_i is the weight of factor i
x_i is the criterion score of factor i

Equation 1 reflects the influence of all combined factors (land-cover, soil type and slope, etc.) and constraints to provide a suitability value to every cell of the data layer [27]. A minimum value of suitability is chosen based on Ordered Weighted Averages using a combination of factors or decision variables.

3.2 Data and Scale

The model uses a synthetic landscape created with the GIS software Idrisi Kilimanjaro [16] to represent the different land-use types of water, undeveloped land, developed land, and urban land. The scale extents for the CA model are both temporal and spatial relating to the specific archeological phenomenon being studied. EARLI uses a temporal resolution of one year for each time-step and a cell resolution of 10 m. Spatially, because settlement dynamics and land-use are both affected by topographic (surface features) and environmental variability [8] a 10m cell resolution was used as it best reflects local scale (i.e., a small dwelling or small plot). A one year time-step is used because most agriculture is based upon a yearly 193 cycle with differences in yearly weather patterns affecting yield rates [13]. This is supported by other ethnographers [19, 37, 39] who report yearly cycles around cultivation and clearing and village building.

3.3 Model Parameters

Yield rate (kg/ha) (Y_{crop})
Every early agricultural complex starts with a small variety of crops it can cultivate based on both local domesticates and foreign varieties. Prominent examples include wheat and barley for the Fertile Crescent, rice for Eastern Asia, maize for North and Central America and taro for Polynesia. Studies of ancient yield rates show them to be 33–50 % less than modern yields [3]. Therefore, the yield rates used are an average for the crop under normal conditions.

Crop type (cal/kg) (C_{kg})
Different crops have different nutritional values. For example, wheat has 1800 calories/kg whereas chickpeas have approximately 1200 calories/kg. Like crop yield sizes, the nutritional values of crops have also increased over the millennia and are therefore adjusted. This value is chosen by either individual crop type or as an average if multiple crops are grown.

Daily Calorie Requirements (C_{req})
Today, even though the recommended daily intake is 2000 calories, individuals in subsistence farming communities generally consume between 1100 and 1800 calories per day [54]. Hence, the actual value used can vary between 1000 and 2500 calories needed per individual adult so as to accommodate impoverished and abundant scenarios.

Reliance (Rl)
Early farming communities subsisted on domesticated plants with additional food resources coming from domesticated animals and supplemental fishing/hunting [24]. The reliance parameter provides a value to indicate this level and is used in calculating the amount of food needed by the community.

Soil nutrition (Sn)
Soil exhaustion is a problem in agriculture, whereby the nutrients deplete through over-use, dramatically affecting yields. The soil nutrient modifier is dependent on local factors and is determined by geomorphology and probable crop types. This parameter is generally calculated for the model to be a multiyear average based on the type of field renewal strategy employed. For example, an intercropping strategy using a combination of cereals and legumes generally has a value of 80 % soil nutrition.

Fallow length ($A_{cultivated}$)
Fallowing of fields is a method of restoring soil nutrients. Two values are used to describe the process as follows: One is the length of time a field remains fallow (F) and the other is the number of years a field is planted (P). The fallow equation is:

$$A_{cultivated} = P\left(\frac{A_{total}}{P+F}\right) \tag{2}$$

where A_{total} is the total amount of developed land and $A_{cultivated}$ is the amount of land available for cultivation. In cases where no fallowing is used $P = 1$ and $F = 0$. Fallowing can increase the amount of land change without increasing the potential population.

Birth spacing (**Bs**)
The birth-rate and spacing of children helps determine how quickly a community can grow. Much of this information is estimated through ethnographic sources [32]. Its value is culturally dependent and is influenced by resource availability. The value selected is for the maximum potential.

Work for clearing new land per Individual adult ($\mathbf{W}_c$)
New fields need to be created by clearing undeveloped land to produce more food for an increasing population. Land clearing removes an area of forest/scrub before tilling and planting. Work required to clear the land is greater than working developed land and should be calculated based on the local environment. For example, in traditional Indonesian villages like Mikol Aloi, which is surrounded by dense forest, an individual could clear 0.5 ha/year using stone axes [37]. This value is between 30 and 50 % of the developed land work value.

Work on developed land per individual adult ($\mathbf{W}_d$)
The calculation of work per adult unit includes such activities as ploughing, sowing, weeding, harvesting and crop-processing. There are very few references in the literature that sets a value to the amount of land worked, in hectares (ha), per adult workers. Lehner [43] whose work cover Pharonic Egypt sets a value at 5 ha/year, while Kirkby [39] sets a value 2 ha/year by hand or 8 ha/year with oxen. Ladefoged [42] in assessing rain-fed and irrigated agriculture of pre-contact Hawai'i has a range of values from 0.6 ha/year on the big island of Hawai'i to 5.8 ha/year on Kauai. The value used during testing was 2.8 ha/year, but should be adjusted to reflect the specific region under study.

Non-workers ($\mathbf{W}_{non}$)
The quantity of developed land is based upon the agricultural work force. The total population and settlement size vary based on the number of non-workers the culture can support and allow. In more egalitarian societies this value is low, 10–20 %, and in more complex and stratified societies this value can be 100 % or even 200 %.

3.4 Model Stochastic Parameters

Weather
Before the advent of irrigation, agriculturalists relied solely on precipitation for watering their crops. In EARLI, the weather is divided into several categories, each having an associated yield modifier. The frequency of each weather type depends on local climatic conditions, but can be broadly classified into two types humid/wet and arid. For example, in a wet/humid environment the frequency of drought is very low (2 % or once every 50 years) whereas in an arid environment the frequency of drought can be every 10 years. The weather element is determined from reconstructed local environmental profiles using paleo-environmental data. The profile data can be derived from fossil pollen, lake sediment, mineral content, diatoms and lake fluctuations [35].

Birth rate, aging, family composition
Families form a basic unit within EARLI and are comprised of 1–4 adults in two groups (adults 15–35 years of age and grandparents 35+ years of age), and 0–12 juveniles in three age groups (infants 0–5 years, children 6–10 years and teens

11–15 years of age). Family growth and aging occurs every fifth time-step with members of each age category moving to the next and new babies being born. As teenagers come of age, the model pairs two new adults to create a new family.

Mortality
During the aging phase a chance of death is calculated stochastically based on mortality curves that follow well defined local mortality profiles using ancient cemetery data [15]. Infant mortality is usually high (30 %) but may be lowered when food reserves are adequate, due to easier access to weaning food.

3.5 Model Variables

Total Yield (**Y_{total}**)
Total yield, which represents available food at each time-step, is one of the key variables that determine the future path of the settlement. To calculate total yield, the total area used for agriculture is multiplied by crop yield rate (kg/ha) and calories per kg divided by daily nutritional requirements using the equation:

$$Y_{total} = A_{cultivated} Y_{crop} C_{kg} \tag{3}$$

where:

$A_{cultivated}$ is land under cultivation
Y_{crop} is the Yield based on an average of the plants sown
C_{kg} is an average of the calories from the plants per kg

Community Needs (**Nd**)
The second key variable is community needs, which is derived from total population and daily caloric intake. The equation is:

$$Nd = C_{req} D_y \text{Pop} \tag{4}$$

where:

C_{req} is the calories needed per day per person
D_y is days in the year (365)
Pop is the total population, both workers and non-workers

Storage
Storage is the amount of agricultural surplus remaining after the community subsistence needs and future seed requirements have been met that can be used in future years to compensate for shortfalls [44, 56]. This is especially important in dry-farming conditions where inter-annual availability of yields may vary 100 % or more [34]. Because losses between harvest and consumption from waste, spoilage, insect damage, etc. occur and have been estimated at 30 % [34, 58], the simulation reduces the left-over yield accordingly.

3.6 Model Transition Rules

Transition rules within EARLI do not change a cell, but provide a potential starting point for agricultural development. These transition rules are derived from the suitability of the cell, as defined by the number of adjacent developed cells, if there are enough working hours available and the distance from the settlement center (5). If cells have equal potential, starting selection is done randomly. The choice between locating the field directly adjacent to the current developed area or a distance up to 100 m away is a product of bifurcation dependent upon community risk level decisions. The equation used in this model is:

$$P_C(t) = \frac{1}{d^{0.2}} * D_C(t) * f(Env_{suit}) * W \tag{5}$$

where:

P_C is the attractiveness of a cell to be the starting point for a field,
t is a discrete time step,
d is distance from settlement center,
W is the available work for clearing new land and is set to either 0 or 1,
D_C is the number of developed cells divided by the number of cells in the neighborhood, and $f(Env_{suit})$ is the suitability of a cell.

4 Model Implementation

The Early Agricultural Resource and Land-use Investigation (EARLI) model was programmed with Python [55], an open-source programming language that uses both object- oriented programming and linear scripting. EARLI is principally a land-use/land-cover change model built around human decisions in response to environmental feedback. The suitability maps were derived in IDRISI using the MCE module. These GIS data layers were applied to the complex systems model framework consisting of: CA to provide potentials of land-use change; agent-based modeling in a stochastic framework for decision making; and discrete time-steps to allow feedback between the model components (Fig. 1).

The potential for land-use change is based on the needs of the community. If the community has enough arable land to feed itself and can only work the existing land, no land change will occur. However, if extra work is available new fields (developed land) will be created. This in turn will increase the potential yield in the next time step. The increase in available food will then feed a growing population. The amount of food that is available for consumption by the community is based upon yearly yields. Yields are calculated both deterministically (soil nutrition and crop yield rates) and stochastically (weather and available land). In parallel, the model determines the community needs based upon total population and daily calorie requirements. If the

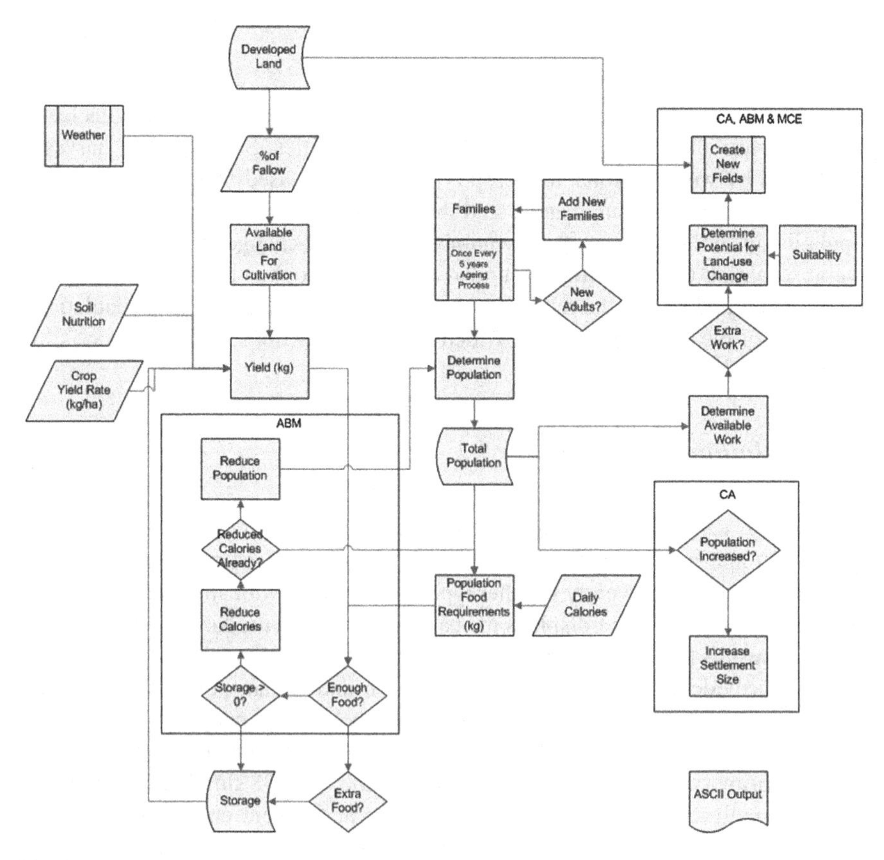

Fig. 1 Flowchart of the EARLI model. The larger outlines indicate which part of the model uses CA, ABM or MCE

yield is enough to provide for the community needs (after seeds are removed), then extra yield is placed in storage. However, when the yield is not enough, the community will use its storage or will reduce its daily caloric intake to accommodate. If there is still not enough, either up to half of the community members will decide to emigrate or individual community members will die until the balance between yield and needs is restored. After this is determined, the amount of available work for cultivation is determined and compared to current available land. Once extra work is available on newly cleared land, the arability then increases. Every five years family demographics are calculated, in which new families are formed, babies born, and some individuals die. If the population increases then the settlement will grow and change developed land to urban land.

5 Results

The central focus of this study was to model farmer's behavioral interactions leading to land-use changes in early agricultural communities. Since ancient harvests are affected by weather, soils, and crop types this study used final yields to determine settlement population changes and thus the amount and type of land needed. Population in settlement size, family numbers, and land-changes were tracked, and several scenarios were investigated under varying environmental and social conditions to better understand the relationship between land-use change and population. The details of the simulations are provided in the next sections.

5.1 Simulation Outcomes

The primary scenarios were those that looked at the difference between *arid* (scenario 1) and *wet* (scenario 2) climates. In these scenarios only the stochastic chance of any particular yearly weather type was adjusted, with all other parameters remaining the same. Table 1 displays the parameter settings used in both scenarios 1 and 2. The results for each scenario are based on 100 simulations of 300 yearly time-steps. This provides a range of probabilities for both land-use and population changes. A simulation will terminate if the population falls below four people.

In the arid environment only 62 of the 100 simulations completed all 300 time-steps. The primary cause for simulation failure in the other 38 simulations was poor weather, usually consecutive years of drought or at frequent enough intervals so the yields from better years could not compensate for numerous food shortfalls. Of the 62 remaining runs, the maximum amount of developed land varied from 16 to 211 ha (Fig. 2a) and the settlement size ranged from smaller than 0.5 to 1.37 ha with between 5 and 26 families representing a maximum population of over 200

Table 1 Parameter settings used in scenario 1 and scenario 2 with ranges used in sensitivity testing

Parameter	Set value	Value range
Land suitability	70	50–100
Starting families	4	2–10
Daily calories	1800	1800–2400
Crop yield (kg/ha)	900	700–1500
Crop nutrition (cals/kg)	1500	1000–2000
Soil nutrient modifier	0.85	0.60–1.00
Reliance on agriculture	0.65	0.50–0.90
Birth spacing (months)	20	15–36
Individual work (ha/yr)	2.8	2.0–5.0
Individual clearing (ha/yr)	0.5	0.5–1.5
Supported non-workers (%)	25 %	0–200 %

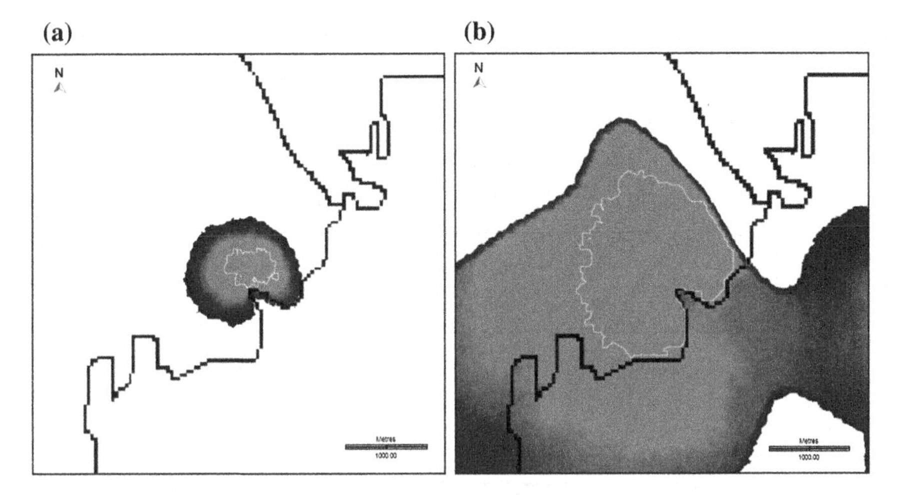

Fig. 2 Spatial extents of all simulations performed in both **a** arid and **b** humid scenarios. The lightest grey areas represent higher possibility of spatial extents that were in all simulations and darker grey areas represent lower possibilities spatial extents of only very few simulations. The internal line represents the minimum amount of developed land

individuals. The majority (66 %) of the completed simulations ranged between 30 and 105 ha of developed land. Within any simulation, a probability of favorable environmental conditions exists, which will represent the greatest extent of both land-change and population. In scenario 1, a single outlier of 211 ha of developed land occurred, due to a positive feedback loop based on a continuance of good weather.

The results from a wet environment indicate a marked difference in every category (Table 2). Of the 100 runs, only 8 did not complete all 300 years and all those failed with the first 10 years due to either too much or too little precipitation. The range of developed land was between 281 and 1472 ha (Fig. 2b), with a majority of the runs having developed land greater than 1000 ha. The settlement size varied from

Table 2 Comparison of developed land areas, settlement sizes, populations and number of families between scenario 1 and scenario 2

	Scenario 1	Scenario 2
Percentage of failed settlements	38 %	8 %
Maximum land developed (ha)	211.9	1472.3
Minimum land developed (ha)	16.0	284.8
Median land developed (ha)	74.5	1210.9
Maximum settlement size (ha)	2.0	21.2
Minimum settlement size (ha)	0.2	2.3
Median settlement size (ha)	0.8	9.3
Maximum population (approx)	200	3000
Mean population (approx)	65	1600
Maximum number of families	26	375

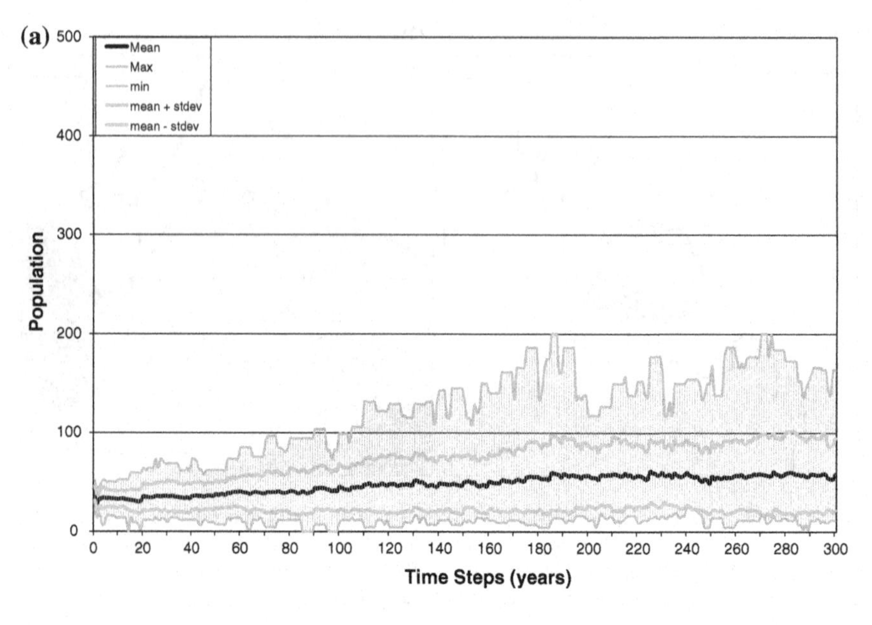

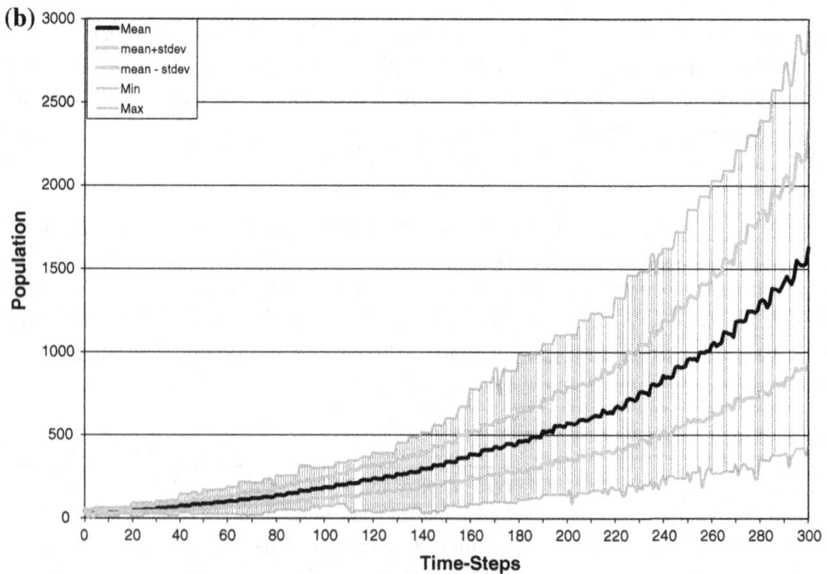

Fig. 3 Population probabilities for **a** arid environments and **b** humid environments

2.2 ha, representing 34 families or a total population of 210 to a settlement size of over 21 ha, representing 375 families and a total population of 3000 individuals. One reason for this is that increased yields from more reliable precipitation help reduce infant mortality, thus providing a more stable and growing population.

Aggregate population curves (Fig. 3) of each scenario depict that in arid environments population rises to a sustainable level then is self-correcting according to environmental conditions. On the other hand, populations in wet environments continue to rise until the maximum sustainable population is reached after all available arable land is used. To examine more closely the influence of increased population rates, another scenario (scenario 3) using a closer birth spacing (15 vs. 20 months) was conducted in the same arid environment. Similar to scenario 1, only 66 % of the simulations completed 300 time-steps. One difference, however, was in the number of runs that could not survive the first 10 years. For scenario 1, 15 of the simulations failed within the first 10 time-steps, whereas in scenario 3, 28 simulations failed in the same first 10 time-steps. However, if the community in scenario 3 could survive these first 10 years they had a better success rate to complete the remaining time-steps. The runs that completed also showed greater land-use change ranging from 44 to 800 ha (Fig. 4). Like scenario 1 the majority of the sites were positively skewed, with a similar distribution of maximum land-use change values around the mean. The settlement size for the majority of runs was 1–4 ha, with a maximum of 8 ha, representing 143 families with over 1100 in population. The results from this scenario depict the importance of steady population increases through increased birth rates which were reflected in the land-use.

To better understand the effects of geomorphology on land use change and population in early farming communities, the starting position of the initial fields was moved against a non-suitable area, thus restricting the direction of development. This scenario (scenario 4) also used the same parameters as scenario 1. The results show that in scenario 4 the number of failures was greater than 50 %, but in simulations

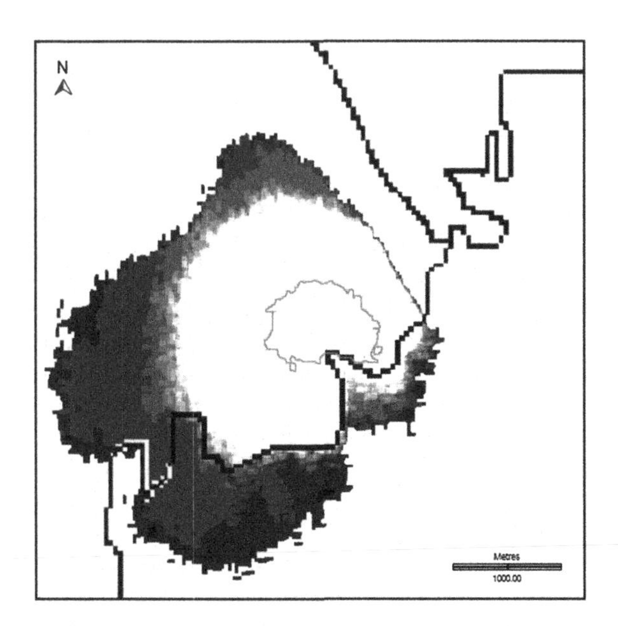

Fig. 4 Spatial extents of Scenario 3 (closer birth spacing). The lightest areas represent higher possibilities spatial extents that were in all simulations and darker areas represent lower possibilities spatial extents of only very few simulations. The internal line represents the minimum amount of developed land

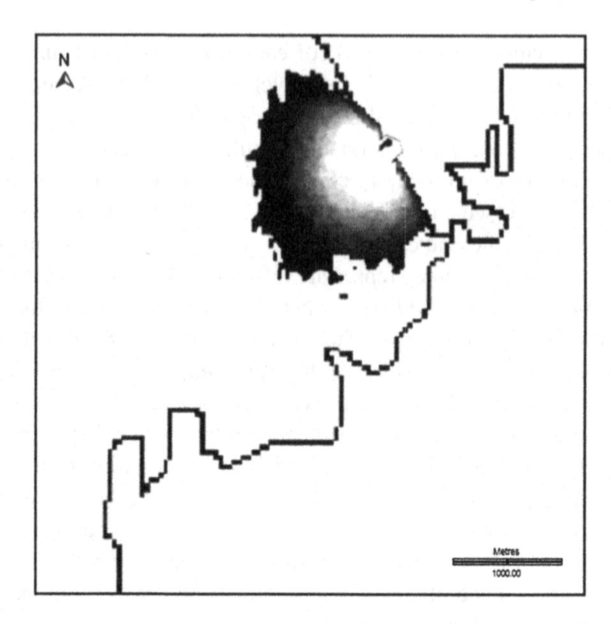

Fig. 5 Spatial extents of Scenario 4 (different starting location). The lightest grey areas represent higher possibilities spatial extents that were in all simulations and darker grey areas represent lower possibilities spatial extents of only very few simulations

that completed the land-use and settlement size ranges were very similar to scenario 1 with the majority between 35 and 100 ha (Fig. 5).

5.2 Sensitivity Analysis Tests

Sensitivity testing of the component parameters has to be a part of sensitivity analysis, examining how the model reacts to small changes in input parameters. While sensitivity analysis can be multivariate, univariate sensitivity tests were performed to determine how changes in a single input variable affects changes in output results and to what degree, without stochastic perturbation [18]. If a parameter has an illogical outcome, adjustments can be made within the model's framework. This can be an iterative process that should be coupled with calibration. For example, the influence of yield was initially predicted to be too low, and was adjusted accordingly.

In each test, the weather parameter was kept the same so that each scenario focused on the parameter in under examination. The sensitivity tests that were performed looked at crop yield rates (scenario 5), crop nutrition values (scenario 6) and soil nutrient levels (scenario 7). The ranges used in each scenario are presented in Table 1.

The results for scenario 5, which examined the effect of differences in crop yield rates indicated land-use changes and settlement sizes increasing as crop yields

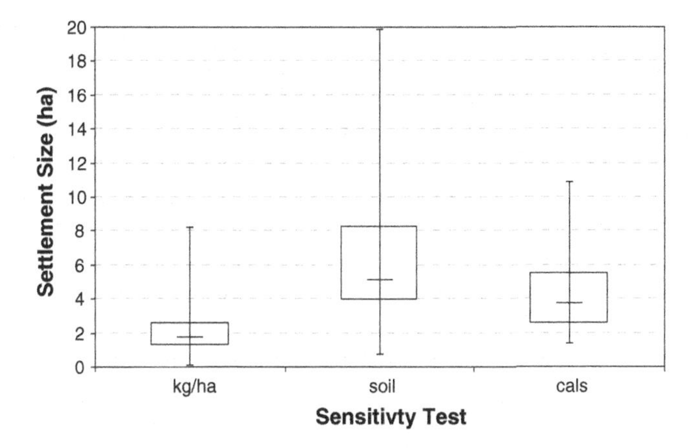

Fig. 6 Box plots of developed **a** land potential and **b** settlement size depicting minimum, 25–75 %, median, and maximum

increased. In scenario 6, increases in calories/kg of different crop types again showed the same trend with subsequent increases in both developed land and settlement size. Both of these results were expected. Soil nutrient results were slightly different. If the soil nutrients were below 60 % no simulation succeeded, whereas once the soil nutrients exceeded 90 % all simulations succeeded. Results further indicated that at soil nutrient levels between 60 and 75 % the difference between minimum and maximum land-use was large (120–844 ha) and at soil nutrient levels higher than 95 % the difference in land-use was less pronounced (942–1393 ha).

Comparing all the results (Fig. 6), the greatest influence on land-use change and settlement size was the soil nutrients parameter. These scenarios also illustrated that no one parameter has an overly high influence on the model's output.

6 Discussion and Conclusion

EARLI model was developed as a theoretical model and as a simulation tool to simulate and analyze the spatial and temporal land-use change patterns and population response of early agricultural communities under a variety of environmental and cultural conditions. The generated simulation scenarios indicate that the EARLI model can produce different land-use change patterns and population growth trajectories. For a given population size the regular zone of cultivation lies somewhere between the maximum and minimum amount of land needed with long term corrections through increasing acreage [58]. The location and size of large settlements can be the result of the surrounding area having very good suitability for cereal crop growth.

The results indicate that sustainability can be tenuous, with maximum potential seldom being attained. This suggests that a complex interplay of weather, crop types, soil nutrition and human decisions over time contribute to determine final outcomes.

The settlement sizes calculated in scenario 1 are similar to settlement size in a known arid environment like in Ethiopia, where most sites are less than 2 ha (D'Andrea personal communication). In scenario 2, the population increases indicated are also seen in Early Polynesia, a humid climate, where population was able to increase quite rapidly [6, 42]. This correlation helps in understanding what range of cultural behaviors may have been employed within a particular environment to create similar demographic and land-use patterns. Studies in Europe have shown that the past agricultural use can be interpreted not only with regard to the natural environment but also with particular societal behaviors [30]. Also, because the archaeological remains are generally the maximum size the village reached, the EARLI model can generate the spatial information on the build up and the decline of a community.

While yield is an important feedback for population increase, birth spacing and mortality can play a constraining role, especially in runaway yield/population growth cycles. McCann [49] suggests that the dominant factor in defining paleodemographics is fertility rather than mortality. The birth spacing parameter can be set by the model users, yet this parameter is only the minimum separation between births. By increasing the birth rate potential from a maximum of three live births per five years to four a more stable population will emerge, which can better cope with adverse climatic/weather effects in the long term. However, as indicated in scenario 3, a higher birth rate is more susceptible to population collapse as agricultural communities are first being established. Other models [25] use population growth rates to predict spatial outcomes, but the stochastic nature of the family life cycle used in EARLI provides better feedback and thus depicts a wider spatial range of land-use change.

The EARLI model uses stochastic elements for weather parameter to generate simulations, while keeping all other parameter constant. For example when the reliance on crops value is set at 65 %, the population remains low but is able to easily adapt to times of stress (e.g., drought). This is important in the first couple of generations as the agricultural area is becoming sustainable. However, after a certain time the reliance level can become a brake because the full potential of the community's labor pool is not used. Conversely, when reliance has a value of 90 %, surviving lean years becomes more difficult and a higher percentage of settlements do not survive. Settlements that do survive the initial growing stages and are sustainable become more productive with higher land-use change and higher population.

The results from testing EARLI in both an arid and humid environment confirm that population and land-use change patterns of early agricultural communities have a range of outcomes that are reflected in the archaeological record. Overall, using a hypothetical landscape, the model has met expectations and therefore can be used to model known archaeological sites with available data. Additional parameters can be added, such as livestock and fodder to better simulate particular communities. Finally, because the model only supports community wide growth patterns and crop selection (i.e. a single crop or an average of several) future development will allow household level decisions for crop selection and birth spacing coupled with improved agent interaction. The future work entails more detailed model validation and comparisons of model results with the actual archaeological records.

Acknowledgments The full support of this study was provided through Social Sciences and Humanities Research Council (SSHRC) research grant awarded to the second author. Authors are thankful for valuable reviewers' comments.

References

1. Amedeo, D., Golledge, R., Stimson, R.: Person Environment Behavior Research: Investigating Activities and Experiences in Spaces and Environments. Guildford Press, New York (2009)
2. An, L., Linderman, M., Qi, J., Shortridge, A., Liu, J.: Exploring complexity in a human-environment system: An agent-based spatial model for multidisciplinary and multiscale integration. Ann. Assoc. Am. Geogr. **95**, 54–79 (2005)
3. Araus, J.L., Slafer, G.A., Romagosa, I., Molist, M.: FOCUS: Estimated wheat yields during the emergence of agriculture based on the carbon isotope discrimination of grains: Evidence from a 10th millennium BP site on the euphrates. J. Archaeol. Sci. **28**, 341–350 (2001)
4. Baltzer, H., Brown, P.W., Kohler, T.: Cellular automata models for. Environ. Plann. B **21**, 531–548 (1998)
5. Batty, M., Xie, Y.: From cells to cities. Environ. Plann. B **21**, S31–S48 (1994)
6. Bellwood, P.: The First Farmers : Origins of Agricultural Societies. Blackwell Pub, Oxford (2005)
7. Benenson, I., Torrens, P.: Geosimulation : Automata-Based Modeling of Urban Phenomena. Wiley, London (2004)
8. Bevan, A., Conolly, J.: GIS, archaeological survey, and landscape archaeology on the Island of Kythera, Greece. J. Field. Archaeol. **29**, 123–138 (2002)
9. Bharwani, S.: Understanding complex behavior and decision making using ethnographic knowledge elicitation tools (KnETs). Soc. Sci. Comput. Rev. **24**, 78–105 (2006)
10. Bogucki, P.: Neolithic dispersals in riverine interior central europe. In: Ammerman, A., Biagi, P. (eds.) The Widening Harvest. Archaeological Institute of America, Boston (2003)
11. Brown, D.G., Riolo, R., Robinson, D.T., North, M., Rand, W.: Spatial process and data models: toward integration of agent-based models and GIS. J. Geog. Syst. **7**, 24–47 (2005)
12. Burnham, B.O.: Markov intertemporal land use simulation model. Southern J. Agr. Econ. **5**, 253–258 (1973)
13. Butler, E.A.: Sustainable agriculture in a harsh environment: An Ethiopian Perspective. In: Hassan, F. (ed.) Droughts, Food and Culture: Ecological Change and Food Security in Africa's Later Prehistory. Kluwer Academic, New York (2002)
14. Carver, S.: Integrating multi-criteria evaluation with geographical information systems. Int. J. Geogr. Inf. Sci. **5**, 321–339 (1991)
15. Chamberlain, A.: Human Remains. British Museum Press, London (1994)
16. Clarke Labs. IDRISI GIS and Image Processing Software. http://www.idrisi.com (2006). Accessed 12 Aug 2006
17. Clarke, D.: Analytical Archaeology. Methuen & Co Ltd, London (1968)
18. Crosetto, M., Tarantola, S., Saltelli, A.: Sensitivity and uncertainty analysis in spatial modeling based on GIS. Agr. Ecosyst. Environ. **81**, 71–79 (2000)
19. D'Andrea, A.C., Lyons, D.E., Haile, M., Butler, E.A.: Ethnoarchaeological approaches to the study of prehistoric agriculture in the ethiopian highlands. In: Van der Veen, M. (ed.) The Exploitation of Plant Resources in Ancient Africa. Kluwer Academic, New York (1999)

20. Davies, A.L.: Upland agriculture and environmental risk: a new model of upland land-use based on high spatial-resolution palynological data from West Affric, NW Scotland. J. Archaeol. Sci. **34**, 2053–2063 (2007)
21. Deadman, P., Brown, R.D., Gimblett, H.R.: Modelling rural residential settlement patterns with cellular automata. J. Environ. Manage. **37**, 147–160 (1993)
22. Dean, J.S., Gumerman, G.J., Epstein, J.M., Axtell, R.L., Swedlund, A.C., Parker, M.T., McCarroll, S.: Understanding anasazi culture change through agent -based modeling. In: Kohler, T., Gumerman, G.J. (eds.) Dynamics in Human and Primate Societies. Oxford University Press, Oxford (2000)
23. Dennel, R.W.: The origin of crop agriculture in europe. In: Cowan, C.W., Watson, P.J. (eds.) The Origins of Agriculture. Smithsonian Institution Press, Washington (1992)
24. Dhavalikar, M.: The First Farmers of the Deccan. Ravish Publishers, Pune (1988)
25. Dickson, D.B.: Ancient agriculture and population at tikal, Guatemala—an application of linear-programming to the simulation of an archaeological problem. Am. Antiquity **45**, 697–712 (1980)
26. Dragićević, S., Marceau, D.J.: A fuzzy set approach for modeling time in GIS. Int. J. Geogr. Inf. Sci. **14**, 225–245 (2000)
27. Eastman, J.R., Jin, W., Kyem, P.A.K., Toledano, J.: Raster procedures for multi-criteria/multi-objective decisions. Photogramm. Eng. Remote Sens. **61**, 539–547 (1995)
28. Fedick, S.L.: Land evaluation and ancient maya land use in the Upper Belize River area, Belize, Central America. Lat. Am. Anitq. **6**, 16–34 (1995)
29. Fischer, M.D.: Culture and indigenous knowledge systems: emergent order and the internal regulation of shared symbolic systems. Cybernet. Sys. **36**, 735–752 (2005)
30. Fliedner, D.: Pre-spanish pueblos in new Mexico. Ann. Assoc. Am. Geogr. **65**, 363–377 (1975)
31. Gregg, S.: Foragers and Farmers: Population Interaction and Agricultural Expansion in Prehistoric Europe. University of Chicago Press, Chicago (1988)
32. Groube, L.: The impact of diseases upon the emergence of agriculture. In: Harris, D.R. (ed.) Origins and Spread of Agriculture and Pastoralism in Eurasia. University College London Press, London (1996)
33. Hoffmann, M., Kelley, H., Evans, T.: Simulating land-cover change in South-Central Indiana:an agent-based model of deforestation and afforestation. In: Jannsen, M.A. (ed.) Complexity and Ecosystem Management: The Theory and Practice of Multi-Agent Systems. Edward Elgar Publication, Cheltenham (2002)
34. Hole, F.: Economic implications of possible storage structures at Tell Ziyadeh, NE Syria. J. Field. Archaeol. **26**, 267–283 (1999)
35. Holl, A.F.C.: Holocene Saharans: An Anthropological Perspective. Continuum, London (2004)
36. Irwin, E.G., Geoghegan, J.: Theory, data, methods: developing spatially explicit economic models of land use change. Agr. Ecosys. Environ. **85**, 7–23 (2001)
37. Izikowitz, K.G.: Lamet: Hill Peasants in French Indochina. AMS Press, New York (1951)
38. Janssen, M.A., Ostrom, E.: Empirically based, agent-based models. Ecol. Soc. **11**, 37 (2006)
39. Kirkby, A.V.T.: The Use of Land and Water Resources in the Past and Present Valley of Oaxaca, Mexico: Prehistory and Human Ecology of the Valley of Oaxaca, vol. 1. University of Michigan Press, Ann Arbor (1973)
40. Kocabas, V., Dragićević, S.: Assessing cellular automata model behaviour using a sensitivity analysis approach. Comput. Environ. Urban **30**, 921–953 (2006)
41. Kuznar, L.A.: High-fidelity computational social science in anthropology: prospects for developing a comparative framework. Soc. Sci. Comput. Rev. **24**, 15–29 (2006)
42. Ladefoged, T., Kirch, P., Gon, S., Chadwick, O., Hartshorn, A., Vitousek, P.: Opportunities and constraints for intensive agriculture in the Hawaiian archipelago prior to European contact. J. Arch. Sci. **36**, 2374–2383 (2009)
43. Lehner, M.: The fractal house of pharaoh: ancient egypt as a complex adaptation system, a trial formulation. In: Kohler T, Gumerman GJ (eds.) Dynamics in Human and Primate Societies. Oxford University Press, Oxford (2000)

44. Levinson, H., Levinson, A.: Origin of grain storage and insect species consuming desiccated food. J. Pest. Sci. **67**, 47–60 (1994)
45. Louwagie, G., Stevenson, C.M., Langohr, R.: The impact of moderate to marginal land suitability on prehistoric agricultural production and models of adaptive strategies for Easter Island (Rapa Nui, Chile). J. Anthro. Archaeol. **25**, 290–317 (2006)
46. Malczewski, J.: GIS and Multicriteria Decision Analysis. Wiley, New York (1999)
47. Manson, S.M.: Bounded rationality in agent-based models: experiments with evolutionary programs. Int. J. Geogr. Inf. Sci. **20**(9), 991–1012 (2006)
48. Matthews, R.: The people and lanscape model (PALM): towards full integration of human decision-making and biophysical simulation models. Ecol. Model. **194**, 329–343 (2006)
49. McCann, J.C.: People of the Plow. Universtiy of Wisconsin Press, Madison (1995)
50. McClure, S., Jochim, M.A., Barton, C.M.: Human behavioral ecology, domestic animals, and land use during the transition to agriculture in valencia, Eastern Spain. In: Kennet, D.J., Winterhalder, B. (eds.) Behavioral Ecology and the Transition to Agriculture. University of California Press, Berkeley (2006)
51. Naroll, R.: Floor area and settlement population. Am. Antiquity **27**, 587–589 (1962)
52. Parker, D.C., Manson, S.M., Janssen, M.A., Hoffmann, M.J., Deadman, P.: Multi-agent systems for the simulation of land-use and land-cover change: a review. Ann. Assoc. Am. Geogr. **93**, 314–337 (2003)
53. Pettersson, M.: Increase of settlement size and population since the inception of agriculture. Nature **186**, 870–872 (1960)
54. Pfister, F., Bader, H., Scheidegger, R., Baccini, P.: Dynamic modelling of resource management for farming systems. Agr. Syst. **86**, 1–28 (2005)
55. Python Software Foundation. Python home. http://www.python.org (2006). Accessed Nov 2012
56. Robinson, W., Schutjer, W.: Agricultural development and demographic change: a generalization of the boserup model. Econ. Dev. Cult. Change **32**, 355–366 (1984)
57. Situngkir, H.: Cultural studies through complexity sciences: beyond postmodern culture without postmodern theorists. In: Complexity and Cultural Studies Conference, Aalborg University, Denmark (2004), 22–24 Jan 2004
58. Smith, M.L.: How ancient agriculturalists managed field fluctuations through crop selection and reliance on wild plants: an example from central India. Econ. Bot. **60**, 39–48 (2006)
59. Steward, J.: The Theory of Culture Change. University of Chicago Press, Chicago (1955)
60. Verburg, P.H., Schot, P.P., Dijst, M.J., Veldkamp, A.: Land use change modelling: current practice and research priorities. GeoJournal **61**, 309–324 (2004)
61. White, R., Engelen, G.: Cellular automata and fractal urban form: a cellular modeling approach to the evolution of urban land-use patterns. Environ. Plann. A **25**, 1175–1199 (1993)
62. White, R., Engelen, G.: High resolution integrated modelling of the spatial dynamics of urban and regional systems. Comp. Env. Urb. Syst. **24**, 383–400 (2000)
63. Whitelaw, T.: The ethnoarchaeology of recent rural settlements and land-use in Northwest Keos. In: Cherry, J.F., Davis, J.L., Mantzourani, E. (eds.) Landscape Archaeology as Long-term History. Institute of Archaeology University of California, Los Angeles (1991)
64. Widdle, A.: The establishment of agricultural communities. In: Widdle, A. (ed.) Neolithic Europe: A survey. Cambridge University Press, Cambridge (1985)
65. Wu, F., Webster, C.J.: Simulation of land development through the integration of cellular automata and multicriteria evaluation. Environ. Plann. B **25**, 103–126 (1998)
66. Zhang, H., Bevan, A., Fuller, D., Yang, Y.: Archaeobotanical and GIS-based approaches to prehistoric agriculture in the upper Ying valley, Henan, China. J. Arch. Sci. **37**, 1480–1489 (2010)

Chapter 10
Oscillatory Dynamics of Urban Hierarchies 900–2000 Vulnerability and Resilience

Douglas R. White and Laurent Tambayong

Abstract We show the fallacies of the Zipfian and power-law views in quantitative analyses of city-size distributions and of the notion that they are superior, simpler, and more universal than q-exponential or Pareto II distributions because of the single-parameter assumption. Both sets of models have two parameters, but the q-exponential captures a scaling coefficient that is fundamental to understanding now the distribution of smaller sizes that is left out by the "cutoff" parameter of Zipfian and power-law. The additional parameter has a fundamental importance in understanding the historical dynamics of partially independent changes in the parameters of urban hierarchies and their interactions with the trading, larger resource bases and conflict networks (internal and external) in which city networks are embedded. We find, for the historical periods from 900 to 2000, combining Chandler and U.N. data, in three different regions of Eurasia, that q, as a small-city distribution parameter, has a time-lagged effect on β, as a measure of only power-law inflection on the top 10 regional cities in China, Europe and Mid-Asia. Taking other time-lag measures into account, we interpret this as showing that trade, economic and conflict or socio-political instability in towns and smaller-cities of urban hierarchy have a greater effect on the health and productivity of larger cities as reflected in Zipfian size distributions, with growth proportional to size. As a means to help the reader understand our modeling efforts, we try to provide foundational intuitions about the basis in "nonextensive" physics used to improve our understanding how q-exponentials are derived specifically for non-equilibrium networks and distributions with long-range process that interconnect different parts, such as trade and transport systems, warfare and interpolity rivalries. The ordinary entropic measure e is "extensive" in that

D. R. White
Institute for Mathematical Behavioral Sciences, University of California at Irvine, Irvine, USA
e-mail: drwhite@uci.edu

L. Tambayong (✉)
California State University at Fullerton, Fullerton, USA
e-mail: ltambayong@fullerton.edu

V. K. Mago and V. Dabbaghian (eds.), *Computational Models of Complex Systems*, Intelligent Systems Reference Library 53, DOI: 10.1007/978-3-319-01285-8_10,
© Springer International Publishing Switzerland 2014

interactions of random effects are additive; the nonextensive generalization of e_q where $e_{q=1} = e$ is ordinary Boltzmann-Gibbs entropy and the concept of multiplicative departures from randomness in the range $1 < q < 2$ helps to relate the physical processes subsumed in urban systems to the underlying foundation of certain Zipfian distributions from contributions to the non-Zipfian properties of the smaller cities in urban hierarchies.

1 Introduction

What are the effects on cities of the larger complex networks in which they are embedded? The coupling of regional urban hierarchies and long-distance trade networks is among many other network-related open problems in historical research. Changes in one region, for example, China, affect changes in others such as Europe, and vice versa. Chandler [9] and other students of historical city sizes ([1, 8, 18, 19, 24] and others) have contributed data that make it possible to compare the multidimensional shapes of city-size distribution curves over time. Focusing on interactions in Eurasia, we measure the multiple dimensions of urban size hierarchies that go beyond power-law fit or Zipfian models, with measures that offer additional benefit in studying the rise and fall of cities and city system in relation to conflicts and wars as well as the historical dynamics of globalization. Our parameter estimates are based on Chandler's database for historical larger-city sizes at 50-year intervals over the last millennium,[1] complemented by overlapping UN population data from 1950 to the present.

Batty [2] refers to our work on instabilities at the level of city systems [37] and notes from our shared database that the top echelon of cities in a single region may be swept away in a short period by interregional competition. This is parallel to our critique of supposedly universal Zipfian shapes of city-size hierarchies [38]. Using different methods, he finds, as do we, that "our results destroy any notion that rank-size scaling is universal… [they] show cities and civilizations rising and falling in size at many times and on many scales." "It is now clear that the evident macro-stability in such distributions" as urban rank-size or Zipfian hierarchies at different times "can mask a volatile and often turbulent micro-dynamics, in which objects can change their position or rank-order rapidly while their aggregate distribution appears quite stable…." He shows the changing ranks of cities ([2] "rank-clock") graphically and charts the micro-dynamics of individual city rise and fall but not the macro-dynamics of rise and fall of city hierarchies in different regions. In general, however:

> "Cities are…emergent, far from equilibrium, requiring enormous energies to maintain themselves, displaying patterns of inequality spawned through agglomeration and intense

[1] From 900 CE to 1970 his size estimates cover over 28 historical periods, usually spaced at 50-year intervals, always comprises a set of largest cities suitable for scaling in a single period. These large stable (and total) cities include 20(80) Chinese, 18(91) European, and 22(~90) Mid-Asian.

competition for space, and saturated flow systems that use capacity in what appear to be barely sustainable but paradoxically resilient systems."[2]

"Where the focus is on interactions between cities in terms of trade or migration ... scaling has recently been discovered[3].... As yet, there are no integrated theories tying these ideas together in a common economic framework consistent with physical scaling, although progress is being made."[4]

"Many objects and events, such as cities, firms and internet hubs, scale with size in the upper tails of their distributions. Despite intense interest in using power laws to characterize such distributions, most analyses have been concerned with observations at a single instant of time, with little analysis of objects or events that change in size through time (notwithstanding some significant exceptions....). It is now clear that the evident macro-stability in such distributions at different times can mask a volatile and often turbulent micro-dynamics, in which objects can change their position or rank-order rapidly while their aggregate distribution appears quite stable. Here I introduce a graphical representation termed the 'rank clock' to examine such dynamics for three distributions: the size of cities in the US from AD 1790, the UK from AD 1901 and the world from 430 BC.[5] Our results destroy any notion that rank-size scaling is universal: "at the micro-level, these clocks show cities and civilizations rising and falling in size at many times and on many scales. The conventional model explaining such scaling on the basis of growth by proportionate effect cannot replicate these micro-dynamics, suggesting that such models and explanations are considerably less general than has hitherto been assumed." [2, p. 592]

Figure 1 (Batty's Fig. 6) tells the Rank Clock story visually, city by city, with colored lines for a sample of nine of the top cities in Chandler's [9] list in each period 430 BCE–1950 CE.

2 Objectives

Our objective is to study the macro dynamics of how separable characteristics of regional city-size distributions are themselves affected by long-range trade, vary over time and affect one another within and between regions with changing urban hierarchical structures: we do not begin with the micro-dynamics of rise and fall of individual cities as in Batty's rank clock. Long-range effects of trade were examined in our previous studies [34, 36] and some will be relevant here at both mid-distance (China and Mid-Asia; Mid-Asia and Europe) and long-distance (China and Europe). White [34] extends the study of spatial effects to include those of peaceful commercial trade at long distances versus forceful trade backed by unequal military-backed exchange that began from Europe in the fifteenth century with the

[2] Batty builds on an earlier working paper [36] on effects of interactions between cities, developed as a part of a common ISCOM project and now published as Chap. 9, In, G. Modelski, T. Devezas and W. Thompson, eds. pp. 190–225. London: Routledge.

[3] As part of the same project ISCOM, see: Bettencourt, L. M. A., José Lobo, Dirk Helbing, Christian Kühnert , and Geoffrey B. West. [7].

[4] Fujita, M. A., A. J. Venables, P. Krugman. [12].

[5] *Prima facie*, because cities appear and disappear, preservation of a Zipfian distributional shape is impossible.

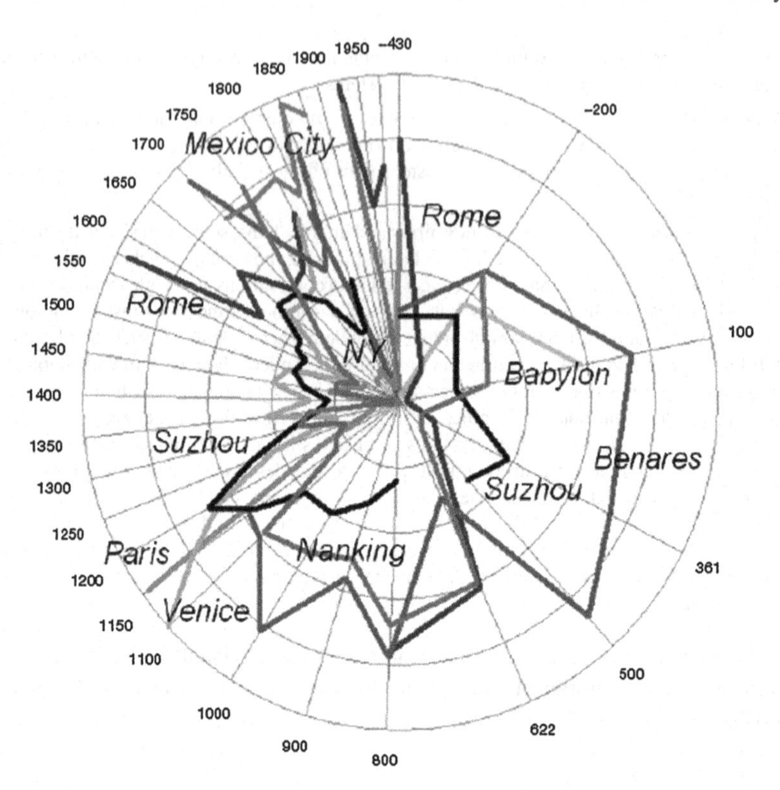

Fig. 1 Rank-clock Trajectory of a sample of world cities (here the radius reflects city sizes) 430 BCE–1950 CE (Adapted from Batty [2])

Portuguese naval invasion of the orient, continued with British and Dutch Imperialism, and then into the twenty-first century with U.S. invasions and industrial colonial globalization. This concurs with Fujita, Venables, and Krugman's [12] work that considers the spatial dimensions of economy at multiple scales as well as Modelski and Thompson's [20] work that combines the interaction between policy-driven competition for economic dominance and political interventions, such as warfare between economic rivals, into a single theory of globalization. Our focus now is on: (1) how cities are affected by political capital (e.g., capitol cities) that serve as international trading centers for regional and long distance trade; and (2) how differences in economically marginal and more central cities in regions interact with trade. To this end, we use the study of city-size hierarchies in order to estimate changes through time in the "dimensions" of urban hierarchies and how they change in relation to one another and to other variables such as trade and war. This is not the same problem as how to simply measure the fit of city size distributions to a power law or related distribution (rank size, lognormal, etc.).

3 Distributions, Parameters, and Questionable Universality

To investigate empirical regularities, as we and Batty have shown from historical data, the focus on the simplest forms of "laws" is not necessarily the most productive approach. Instead, we propose to question the potential interactions among various city-hierarchy variables along with those involving trade and war. In doing so, we propose that the appearances of universality and indefinite self-organization in widely-accepted power-law urban hierarchies may be deceptive, masked by the fact that power-law distribution measures (Pareto I) [23] are often said to have a single parameter:

$$\mathrm{P}(X \geq x) = x^{-\beta}$$

where β is the slope of the line in a log-log biplot that best fits a (linear) distribution of city sizes on the axes x = size and y = frequency (or cumulative frequency of that size or greater). The Zipfian is the special case of $\beta = 2$.

There are in fact two parameters to Pareto I and the Zipfian, not one. Using only one-parameter city-size power-law fitting, one screens off the lower part of the empirical distribution such that the fitting is simplified and biased toward the upper part of the distribution. The second parameter needed is a "chosen" size below which the regression line is no longer power law and the Zipfian no longer holds. The Zipfian is a special case where the power law equation $y = ax^c$ gives a number of cities relative to a given size x an equal population above and below a multiple and divisor c on the x axis.[6]

If we define a city as a dense built-up (urban) area with suburbs but without farmland inside (as does Chandler) with a minimum of 3,000 people (or as does Modelski [18]: "a community with a significant degree of division of labor that makes it a part of a network of cities" [not a settlement of farmers]), then the Pareto I and Pareto II both require two parameters. Neither is more parsimonious than the other. Many studies note that scaling results also depend on the definition of urban boundaries [6]. Not two but perhaps even three parameters are entailed by the fitted power-law for the tails of these distributions: an implicit upper bound beyond which further growth is not sustainable and would produce a phase transition. Larger city sizes may be reached where further growth rate declines. The full extent of city size distributions, however, is often lognormal where rate of growth is roughly constant relative to size (Gibrat's lognormal rule of proportionate growth). Thus for towns or small cities, there is a size at which rate of growth may be relatively fixed (Gibrat's constant) in contrast to the self-similar growth of thinner "fat tails"

[6] Zipfian rank-size for cities ranked 1 to n in size is also equivalent to the tendency to approximate a size of M/r, where r is a city's rank compared to the largest city and M is a maximum city size that best fits the entire distribution. This formulation allows the rank 1 largest city size S1 to differ from its expected value under a Zipfian fitted to an extensive set of the larger cities.

where larger cities may attract immigrants proportionally to size.[7] As a note, the "fat tail" of Gibrat's lognormal distribution is often indistinguishable from a power law, resulting in the aforementioned bias.

4 Interactions Among Measurements for City Size Distributions

For our purposes, we thus need scaling measures for full city size distributions (the whole curve), whereas the Zipfian and Pareto I (power-law) apply only to larger cities (the tail). We use the Pareto II rather than Pareto I to derive more useful city-size distribution measures.

$$P^{\ominus,\sigma}(X \geq x) = (1 + x/\sigma)^{-\ominus}$$

The two parameters $\ominus$ and σ of Pareto II reflect: (1) the rate at which a cumulative distribution, which at low sizes has a little or no tendency to form a power law, begins to asymptote toward a power-law tail; and (2) the slope of the asymptotic limit towards a power-law tail. However, we will present a fuller explanation of an equivalent way to derive this distribution where these two parameters are more evident: the q-exponential with q defining the asymptotic power-law slope $1/(1-q)$ and κ (kappa).the rate of approach.

For $\theta = 2$ and $q = 1.5$, changes in these parameters are locally correlated. Pareto II with log-log slope θ is equivalent to the q-exponential with $q > 1$, where $q = 1 + 1/\theta$, and to the Zipf-Mandelbrot. Each generalizes an inverse power law that depicts thinner tails and asymptotic power law behavior in the tail contrasted with curvature in the body. Thus, the q-exponential function $e_q{}^{x}$ [28, 29], pp. 5–6) covers the full distribution of urban sizes using the theory of Boltzmann-Gibbs entropy.[8] The range $1 < q < 2$ is defined over a cumulative positive distribution which in this chapter we use as that of city sizes. The q-exponential is one of the more developed theories in "behavioral statistical physics" [11]. One way to estimate the q-exponential for such a cumulative distribution is to minimize the linear fit of the natural log $\ln_q(x/\kappa)$ of x/κ, where κ is the multiplicative scale parameter, measuring the degree to which the distribution becomes more multiplicative in the tail (growth as a function of size: more attracting more), and q is the asymptotic limit of slope in the tail of the distribution. The q exponential, then, generalizes the extent and rate at which a distribution approaches a power-law. It can be derived from the following definitions, generalizing BG entropy to include the emergence of power-law distributions:

---

[7] See Clemente and Gonzalez-Val [10].

[8] $q = 1$ is simply the standard free energy distribution of Boltzmann-Gibbs, $e_{q=1=BG}{}^{x}$, analogous to a cumulative distribution function (CDF) over a process with a Boltzmann constant k and in which independent events occur at a constant average rate, except that in $e_{BG}{}^{x=W} = k \ln W$ the average of equiprobable events in the interaction space W is *logarithmic*. Thus BG entropies are *additive* but $e_q{}^{x}$ for $q > 1$ are not.

$$\ln_q(x) \equiv (x^{1-q} - 1)/(1 - q)$$
$$\ln_q(x/\kappa) \equiv \ln_q(x) - \ln_q(\kappa) \equiv ((x/\kappa)^{1-q} - 1)/(1 - q),$$
$$\text{with inverse } e_{x/\kappa}{}^{q} \equiv [1 + (1 - q)x/\kappa]^{1/(1-q)} \equiv \ln_q(x/\kappa)^{-1}$$

The fitted total population $r(x)$ is a function of a population size for a cumulative distribution of population x, the distribution asymptote q, its curvature rate k, and a normalizing constant N_o. Thus, $r(x) = N_o q/\kappa e_q{}^{x/\kappa}$. The q-exponential fitting for a total urban population of Brazil is shown in Fig. 2. This example solves $\ln_q[r(x)]$ for the κ that gives a linear mono-log plot and shows a "king" effect of two outsize primate (largest) cities, Sao Paolo and Rio de Janeiro.

Shalizi [27] shows the equivalence between the q-exponential $e_q{}^{x/\kappa}$ and the Pareto II distribution with parameters theta and sigma $\theta = 1/(q - 1)$ and $\sigma = \kappa\theta$ that can be renormalized to those of the q-exponential (q and κ). His R script gives a Maximal Likelihood Estimation, which is used in fitting the parameters of distributional models of small samples. MLE estimates are unbiased, meaning that the expected values of the estimated parameters for each sample would converge to the true parameter values for all n independent samples of the same data. $P(X \geq x)$ is the probability that an urbanite will reside in a city of at least size x, capturing the complete (Pareto II) shape of the empirical $P(X \geq x)$ as a cumulative complementary distribution function (CCDF).

Figure 3 fits the city-size data for 900–1970 (ca. 23 time periods) cities in Batty's Rank-clock using Shalizi's Pareto II. The scale on the y axis is cumulative probability. This controls for total population in each period, making evident the differences in

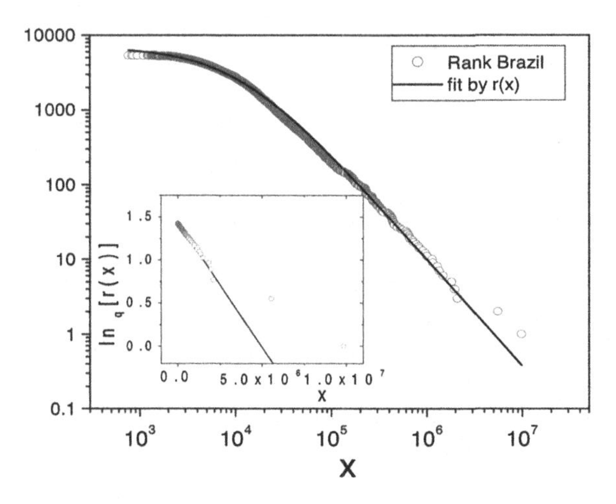

Fig. 2 Fit of cumulative distribution for all cities in Brazil, with $r(x)$ as y-axis. The parameters are $q = 1.7$, $r_0 = 6968.6$ and $a = 0.00024$. *Inset Plot* generalized mono-log plot for Brazilian cities. The coefficient of determination in nonlinear fit (mono-log plot) is $R^2 = 0.99$. From Malacarne et al. [17, p. 2]

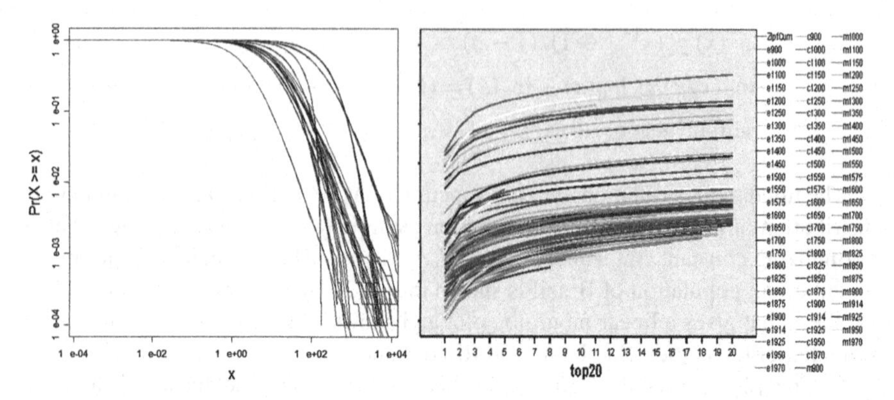

Fig. 3 Fit of cumulative distribution for all Chandler city distributions, 900–1970 CE. *Left* Actual versus Fitted q-exponential distributions for years, Shalizi Pareto II. *Right* Actual versus Fitted q-exponential distributions for selected years

the q-shapes (blue lines) of the distributions both in the body and the heavy tails that tend to be asymptotic power-law. This figure also shows that primate cities in the raw data (black lines) deviate to larger sizes from their otherwise smooth curvatures.[9] The x-axis here is in 1000s times a natural log scale so a smallest city of 4000 would be just below $e^{02} = 4500$. There is an implicit population cutoff of 10,000 at $e^{03} = 9500$ for the smallest city sizes recorded by Chandler so the leftmost parts of the fitted distributions (in black) may be ignored.

5 Hypotheses and Measures

Our theoretical expectation is that the Tsallis entropy model (Pareto II) and Zipf-Mandelbrot distributions will fit city size distributions better than the Pareto I and the Zipfian because of long-range inter-city correlations that operate through trade networks, attraction of migrants and interregional competition and conflict. For Modelski [18]: a city is "a community with a significant degree of division of labor that makes it a part of a network of cities". These effects are mild for small towns and cities, and reach power-law growth proportional to sizes only at larger sizes.

For our purposes, the estimated parameter q adds a measure of the linear slope of $\ln_q([r(x)])$, as in Fig. 2 for Brazil, which reflects the curvature of the overall r(x) in the whole distribution for the entire city-size distribution, and not just that part of the thinner "fat tail" that fits a power-law. We also compute the linear Pareto I slope

[9] Another parameter for each curve (derived from θ and σ of Pareto II) is the value of x where the line on the upper left of each graph as in Fig. 2 intersect with asymptotic power-law line of the tail. We do not use this cutoff for the Chandler data because the lowest city sizes are missing and the top of the graph is truncated. Bercher and Vignat's [5] references [29,37] list standard statistical derivations for Pareto II, including those [32,34] that give MLE results, including Shalizi's.

β for only the top ten cities which in the case of Brazil will show an exaggeration of the "fat tail" for the top two cities that is not reflected in the value of q. We can thus compare our measures with the notion that the Zipfian distribution is not only recurrent but possibly a universal pattern for city sizes as well as many other complex system phenomena.

We use these two estimates—different body and the tail coefficients at 50-year intervals of city-size distributions for different world regions—to show how changes in urban systems relate to general population dynamics and the effects of networks of exchange (we can also use these methods to estimate minimum size estimates for cities and to make fairly accurate historical estimates of rural/urban percentages). For Eurasia and regions within it, we also expect to find that there are systematic deviations from the Zipfian in some historical periods, showing the characteristics of a regional collapse of city systems. True, there is often eventual recovery. Yet, there are also capital cities that never recovered such as Angkor Wat in Southeast Asia and cities in the Mayan civilization.

While noting deviant patterns for larger or primate cities, our approach here is to divide up the Eurasian component of Chandler's [9] largest city-sizes data into 3 large regions—China, Europe, and the Mid-Asian regions in between—and measure two different types of departures: one for the tail (Pareto I slope β for the top ten cities, often power-law but which may or may not be Zipfian) and one for the whole body of the distribution (Pareto II curvature q for the whole body of the distribution). We use measures on each of these standard continuous functions with parameter fitting, based on recasting the data as a cumulative probability in its complementary form.[10]

6 Regional Results for the Three Eurasian Regions

The descriptive statistics for q and β are shown in Table 1. Pareto I and II distributions are said to be consistent when $q = 1 + 1/\beta$ and Zipf-consistent when $q = 1.5$ and $\beta = 2$. $q = 1.5$ corresponds to the whole body of the curve distribution that converges to a Zipfian tail and $\beta = 2$ corresponds to best fit for the 10 top cities (upper tail) that is Zipfian. The historical norm for both tails and bodies of city size distributions from Chandler's data in our period of inquiry (Table 1, upper and middle part) does approximate $q = 1.5$ and $\beta = 2$ for China while mean values for Europe and MidAsia are lower. MLEq values in Tables 1, 2, 3 and 4 simply denote that the estimation of q is done by Shalizi's Pareto II algorithm. Table 4 simply shows that the city data on China is more consistent in factor-analytic variance than the other two regions. q and β are likely to be better approximated in the more recent early modern and modern period where data accuracy is likely to be better.

[10] See Appendix A for the calculation and explanation of these 2 parameters. Bootstrap estimates of the standard error and confidence limits of the q, κ parameters derived from Θ, σ are provided by Shalizi's [27] R program for MLE.

Table 1 Descriptive statistics

	N	Minimum	Maximum	Mean	Std. deviation	Std. deviation/Mean
MLEqChinaExtrap	25	0.56	1.81	1.5120	0.25475	0.16849
MLEqEuropeExtrap	23	1.02	1.89	1.4637	0.19358	0.13225
MLEqMidAsia	25	1.00	1.72	1.4300	0.16763	0.11722
BetaTop10China	23	1.23	2.59	1.9744	0.35334	0.17896
BetaTop10Eur	23	1.33	2.33	1.6971	0.27679	0.16310
BetaTop10MidAsia	25	1.09	2.86	1.7022	0.35392	0.20792
MinQ_BetaChina	25	0.37	1.16	0.9645	0.16247	0.16845
MinQ_BetaEurope	23	0.68	1.26	0.9049	0.15178	0.16773
MinQ_BetaMidAsia	25	0.54	1.01	0.8252	0.13217	0.16017

Table 2 Runs tests at medians across all three regions

	MLE-q	Beta10	Min(q/1.5, Beta/2)
Test value(a)	1.51	1.79	0.88
Cases < Test value	35	36	35
Cases ≥ Test value	36	37	38
Total cases	71	73	73
Number of runs	20	22	22
Z	−3.944	−3.653	−3.645
Asymp. Sig. (2-tailed)	0.0001	0.0003	0.0003

Table 3 Runs test for temporal variations of q in the three regions

	mle_Europe	mle_MidAsia	mle_China
Test value(a)	1.43	1.45	1.59
Cases < Test value	9	11	10
Cases ≥ Test Value	9	11	12
Total cases	18	22	22
Number of runs	4	7	7
Z	−2.673	−1.966	−1.943
Asymp. Sig. (2-tailed)	0.008	0.049	0.052

Table 4 Communalities

	Initial	Extraction
MLEChina	1.000	0.660
MLEEurope	1.000	0.444
MLEMidAsia	1.000	0.318

Because q and β are historically not fully Zipf-consistent for city size distributions (with $q = 1.5$ and $\beta = 2$), we compare the measures of β only for the ten largest cities and q for curvature over in the whole body of the distribution (in Chandler's data, down to a size limit of 10,000). By these criteria we can test for their mutual effects over time and interactions with other variables. **High** $\boldsymbol{\beta}$

(above 2, thicker tailed) means economic boom for largest/primate cities (N = 10), which is commonly associated with thriving international trade and capital cities. However, $\beta > 2.3$ indicates large and dominant primate cities that may be overgrown through trade and attraction of migrants. This could potentially result in their crash if they can no longer sustain themselves. **Low β** (below 2, thinner or "fat" tail) means primate cities (N = 10) are depressed below the expected Zipfian, which is rare. $\beta < 1.7$ often corresponds to destruction of primate cities due to trade competition, war or natural disaster. **High q** (above 1.5) means a city system with thriving medium-large cities associated with largest cities above the Zipfian norm, whether or not there is actually high β which corresponds to thriving primate cities (e.g., smaller cities could thrive through a healthy rural economy somewhat independently of largest city sizes, as with China in the tenth century). $q > 1.8$ often corresponds to an abnormal growth of cities. **Low q** (below 1.5) means a generally depressed city system with small cities that are commensurate with depressed Zipfian tails (low β). When $q < 1.3$, the city system may be approaching a crash, but may not actually be crashing. When $0 < q < 1$, it indicates anomalous system collapse. When $q = 1$, it is a special rare case when a system lacks complex interactions and power-law tails (infinite Pareto II slope), an entropic system "at rest".

The Zipf-normalized minimum (MinQ_Beta) of q and β in lower part of Table 1 is shown for each period and region in Fig. 4 (lower light line). Here we normalize q and β and divide them by 1.5 and 2.0 to take the minimum as a measure of the "worst state" of potential city hierarchy depression, either in q (the whole system) or β (the largest 10 cities). These values are often below a Zipf-norm of 1. In all three of these measures, there is much variation over time. MinQ_Beta values are frequently depressed for MidAsia, and less so for China and Europe.

Crucial to our analysis are statistical runs tests as to whether the variations around means are random or patterned into larger temporal periods. The runs tests in Tables 2 and 3 reject the null hypothesis at $p < 0.00003$ overall, $p < 0.01$ for Europe, $p < 0.05$ for Mid-Asia, and $p < 0.06$ for China.

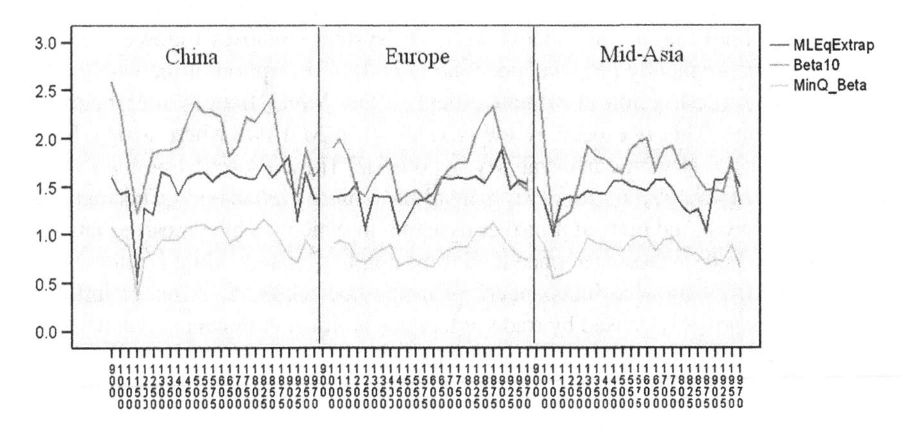

Fig. 4 Values of q, β, and their normalized minimum. Each regional time series runs from 900 CE to 1970

Figure 4 shows the q and β slope parameters fitted by MLE for the China, Europe and the region we call Mid-Asia (from the Middle East to India). We expected the latter to be more unstable due to long-range trade between East and West and buffeted by wars on conquests at both ends and internally. This was borne out by the normalized MinQ_Beta value for MidAsia that is 1.33 standard deviations below the Zipfian value (means test $p < 0.001$, $N = 22$). Table 1, however, shows that the standard deviations for the other 2 regions are even higher than for MidAsia.

Mean and standard deviation values for q in the three regions vary around $q = 1.5 \pm 0.07$, consistent with a Zipfian tail on average, and similarly for variations around $\beta = 2 \pm 0.07$ for the Pareto slope of top 10 cities. Relatively long city-slumps (low q and/or β) occur in the medieval period (1000–1100 for China, 1100 for Mid-Asia, and 1250 for Europe) for all three regions. In 1900, China experiences a sharp drop in q (approaching 1) following the Unequal Treaties of the Opium Wars as the start of China's "Century of Humiliation" at the hands of the British, while Europe experiences an abnormal rise in q beyond 1.7. These depict how the boom in Europe was at the expense of a city system crisis in China. There were also slumps in large-cities occur in Europe in 1400–1500, Mid-Asia in 1800–1875, and China in 1925 after the fall of the Qing when q falls to 1.2.

Historical data show that the slope of the strictly power-law tail of the size distribution (β) tends to generally change faster than the shape body of the distributions (q). This is tested using data from all three regions using the autocorrelation function (ACF), where values of a variable in one time period are correlated in value successive time lags (each lag in this case adding 50 years). The upper and lower confidence limits are at 95 % for a two-tailed significance test ($p < 0.05$). The ACF of β compared to lags of q shows a short-term continuity effect (1 lag of 50 years) and a recovery period at 5–6 lags; then autocorrelation largely disappears. q has more stable long term autocorrelations with itself (up to 16 lags or 800 years). The ratio of q/β has autocorrelation only for Europe, which is oscillatory, but is not significant. If each distribution were Zipf-consistent, q and β would covary perfectly, which they do not.

The occasional very low Zipf-normalized minimum (MinQ_Beta) values in Fig. 4 (lower right line) may be associated with city system collapses following major conflicts, which typically produce a decline in both q (disruption of the whole city system) and β (destruction of primate cities) so that MinQ_Beta is often below a Zipfian average. This is especially the case for β in MidAsia where MinQ_Beta is lowest, consistent with our instability hypothesis. The q/β oscillation involves a tradeoff over time between a more Zipfian tail and a more Zipfian body. This tradeoff is dominated by β and hints at a further dynamic in which we hypothesize that: (1) major dips in both measures at once tend to be conflict-related; while (2) abnormal rises only in β, somewhat independent of q, may be related to crises of inflated primate cities possibly caused by trade or tenuous political dominance such that the cities are in danger of no longer sustaining themselves.

7 Goodness of Fit Versus Possible Inaccuracies in Time Series

Figures 5 and 6 show the departure between raw data from Chandler and fitted values for q and those for β for each data point and interpolation between them. These differences are relatively small throughout for both China (red colors) and Europe (black) except for the following. The tendencies of q and β to diverge in the period 1450–1600 are clearly a systemic interaction between q and β in both figures. In a study of Eurasian globalization, 900–2000, White [35] shows how this period is one of conflict due to the expansion of European domination into Asia where trade was led by military force. For China, divergences are also temporally clustered, with red dots for q data tending below the fitted lines in red when those trends are downward, and tending above the fitted lines in red when these trends are upward. This tendency is muted for β. For Europe the convergence is much greater. This might suggest that the European city size data are more accurate. Clearly, however, the data and the estimated parameters are systematic and not simply a result of biased data. As our theory predicts, systemic patterns of the oscillations of q and β are evident and not likely to be due to differences in data-bias through time.

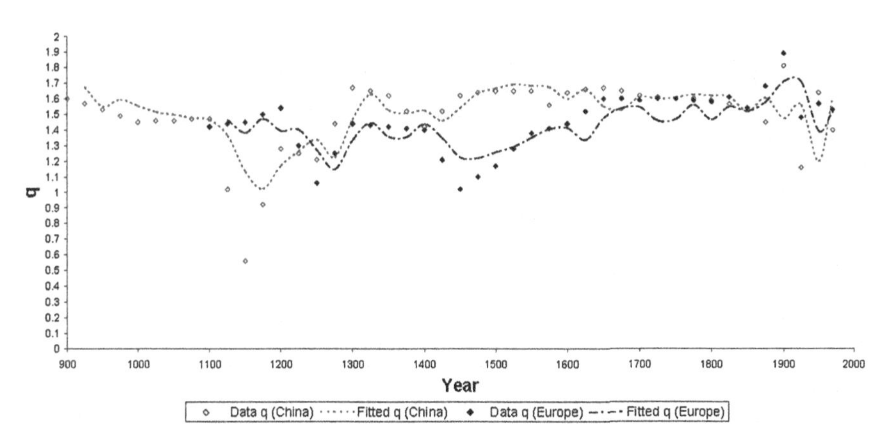

Fig. 5 Time-series of data q and fitted q

More systemically, changes in one value often precede changes in the other. Song China β drops from overly fat tails (2.6) to Zipfian (2.0) to thin (1.4) from 900–1150 as it passes from a strong rural trade economy to a more controlling Imperial city in relocating from Bianjing (Kaifeng) to Lin'an (Hangzhou) and the denser population in the south, while the q body of the distribution remains at a healthy Zipfian (1.6–1.4) until 1100 and then drops precipitously when Imperial city (Lin'an) currency inflation crashes the silk economy. Europe's q, with massive sales of bullion to pay for Silk Trade, remains below Zipfian for a century, and but crashes twice. Similar stories can be told for other periods. Overall, as shown in Fig. 7, fitted values

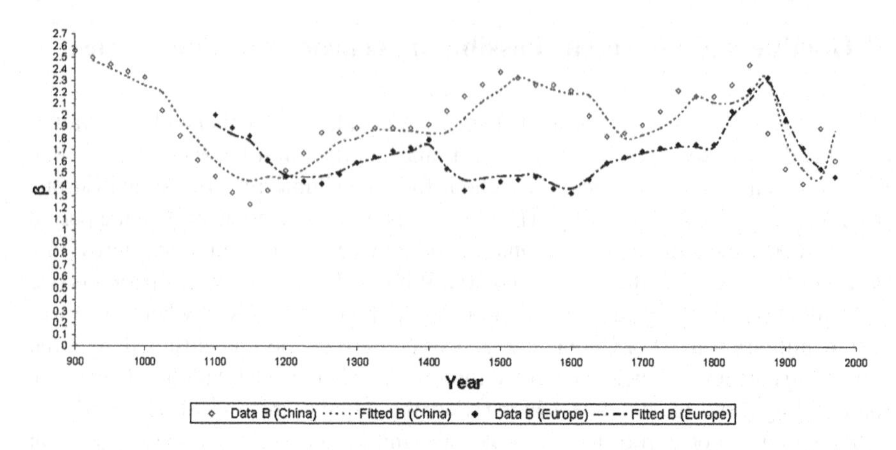

Fig. 6 Time-series for China (red) and Europe (black) showing fitted and actual datapoints

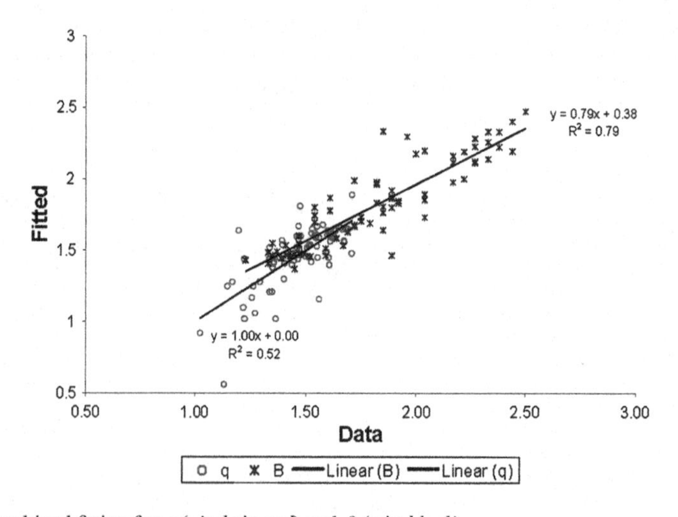

Fig. 7 Combined fitting for q (*circle* in *red*) and β (*x* in *black*)

q and β covary but have high regions of high variance, especially at low values of q (1–1.5) and high values of β (1.8–2.2), both for China and Europe.

8 Interactions of q and β with Cycles of Trade, Conflict, and Changes in City Hierarchies

Recall that a rise in β means a disproportional increase in the size of largest cities. A fall in β indexes a relative decline for large cities, often caused by people leaving the largest cities to go to smaller ones due to war or disaster or its failure to support

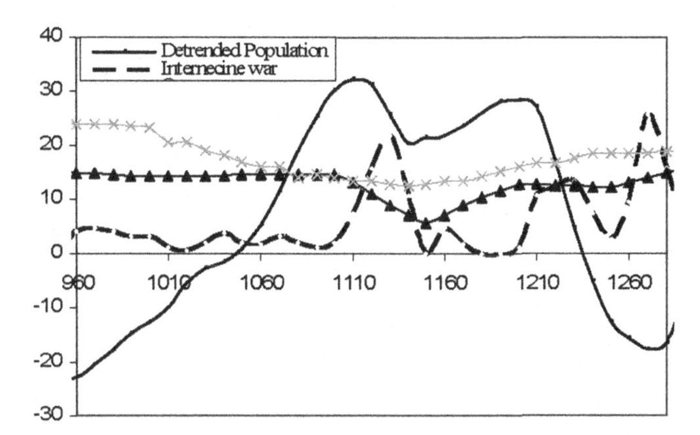

Fig. 8 China's Song Dynasty structural demographic dynamics (960–1279 CE) of negative feedback between population pressure P (*solid curve*, a Turchin [32] measure of detrended population pressure defined by dividing population numbers by net grain yields as an estimate of carrying capacity) and sociopolitical instability (broken SPI curve for internecine wars, J.S. Lee [16]). The changes in β (with x in the lines, starting with slow "fat tail" decline of the large city-size distribution) have a slight time-lagged effect on q (line with *triangles*) after 1110

immigrants. This tends to occur with rising population pressure on resources measured with population growth detrended. Decline in q signals a diminution of smaller cities sizes and works to their disadvantage in indexing disproportional change toward thinner tails (β) with a less proportional growth of large cities.

Decline in β often co-occurs with the peak of a crisis of sociopolitical instability (SPI) or violence that follows a population peak, as exemplified in Fig. 8. Heightened SPI often has to do with wars, like the Song Dynasty loss of their first capital to the Jin in 1127, a date at the center of Fig. 8. The depression of values of q (line with triangles in Fig. 8) occurs once the SPI crisis (elevated dashed line) has ended, followed by rising q with detrended population pressure on resources (solid curve, measured by P, as in Fig. 8), then leveling of q before the next population peak.

As in the endogenous dynamics of sustained structural demographic fluctuation in the two-equation model of Turchin [32], upward detrended population P leads upward SPI (S, Socio-Political Instability) by a generation and downward SPI leads upward P by a generation. Turchin's equations, equipped with appropriate constants, and lagged temporal units at 25 year or generational intervals, are

$$St + 1 \leftarrow Pt \tag{1}$$

$$Pt + 1 \leftarrow -St \tag{2}$$

It is the difference in sign of these reciprocal equations that makes for negative feedback oscillations. Turchin [30, 31] argues that a two-equation oscillatory dynamic works optimally when one of the interactive variables is offset by 1/4 cycle. This tends to be the case with population/resource ratio scarcity (P) measures and sociopo-

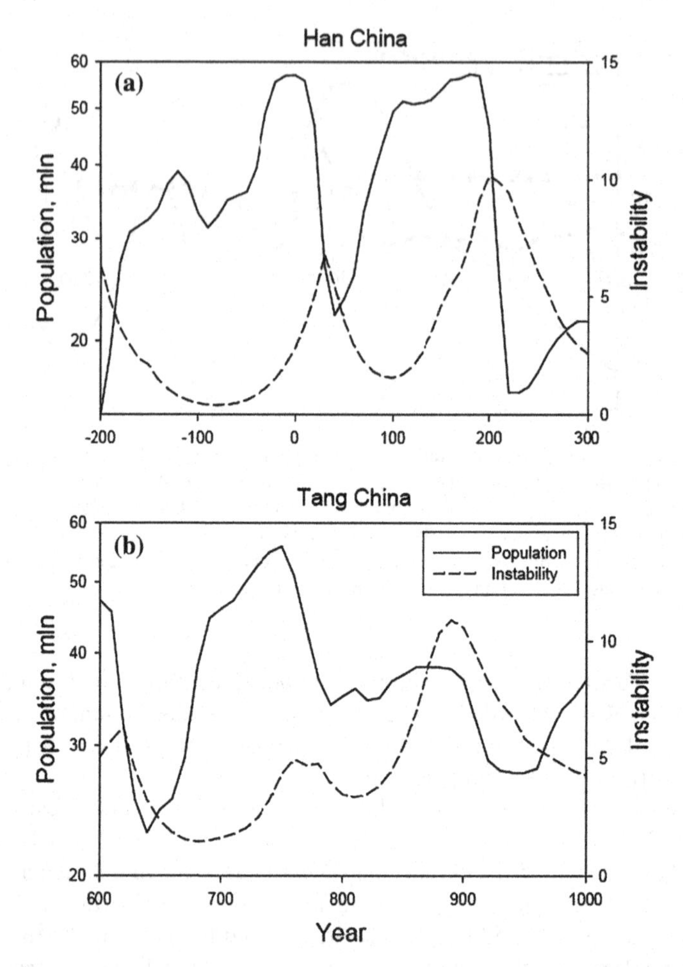

Fig. 9 Dynamics of population (*solid curves*) and sociopolitical instability (*broken curves*) in China. For: **a** Han (200 BCE–300 CE) and **b** Tang (600–1000 CE) periods. From Turchin (2005), with population detrended by bushels of grain

litical violence (SPI or I), as in both Fig. 8 and in fluctuations of the same variables in the previous Han and Tang period of China, as shown by Turchin [32, p. 14] and in Fig. 9a, b.

Also significant is that the three agrarian empire periods (Han, Tang and Song) in Figs. 8 and 9 are separated by periods of external wars and instabilities that interrupt the endogenous dynamics between population pressure and SPI fluctuations. Each period resumes fluctuation modeled by the two-equation time-lagged dynamics of P and SPI, which tends to obtain only when major external disturbances are absent. The external wars in Lee's [16] analysis show up as large SPI fluctuations between dynasties (usually leading to their termination and a later successor empire), while internal

SPI fluctuations occur within the periods of relatively more endogeneity. These various cycles couple into larger cycles of multiple successive empires, observed by Lee [16], as long as 800 years, marked by the most violent of transitions.

Figure 9a, b shows the coupling of two successive "secular cycles" studied by Turchin [30, 32]. Between these two pairs of secular cycles are "exogenous" conflict events, with the second cycle in the second pair (Tang Empire) continuing on in the 960–1000 period into the Song period shown in Fig. 8. The temporal connection of the two disconnected two-phase Han and Tang cycles and succeeding two-phase Song cycle exemplifies structural demographic changes on the order of six irregular 220 ± 40 year cycles (in this case, circa 1500 years total with cycle lengths that are variable). Hierarchical embeddings of irregular oscillatory positive or negative feedback cycles may be stacked such that each has an different exogenous effect on the next. Subject to further study, the negative feedback cycles may have more regular oscillations. The positive ones may have a tendency to move toward equilibrium. It may be the endogenous cycles (e.g., [30, 32, 33]) with negative feedback that are more likely to settle on dynamical offsets of their interactive cycles of 1/4 the cycle length. If there are longer average cycle lengths for endogenous negative feedbacks, longer than those of structural demographics with offset between 1/4 and 1/8 of a cycle, we might find that the time lags are not generational but processes that take longer to have their effects. And because city systems exhibit a positive feedback cycle, their cycle length may be less regular, and may switch between faster and slower time-scales.

Because of spatial or network boundary effects, these kinds of couplings seem to be a more general historical phenomenon and raise the possibility that conflict events within a cycle may kick off oscillations that take twice a long to return to their original state. Turchin cycles with double negative conflict phases and changing effects at spatial boundaries are thought by Modelski and Thompson [20] to embed two leading polity cycles that average about 110 ± 20 years. Our city-dynamics data suggest similar patterns. Averages and timings vary, but these average cycle-lengths might tend to diminish by half as each embedded endogenous process: negative feedback cycles tend to operate at successively smaller spatial scales. We hypothesize an embedding of dynamical processes that run from trading zone network sizes and city-size distributions that cycle roughly 200, 400, or 800 years, partly dependent on the severity of the declines.

A Turchin cycle might take 220 ± 40 years in which time a period of trade expansion might end with warfare and restricted regional boundary conditions as the results of SPI and city system disruption. The next Turchin cycle for this region might operate within these restricted trade boundaries that also run up the cost of defensive perimeters. The conflict phase at the end of that cycle might result in turn in the removal of those defenses, opening into a new period with an expanded trade network. Thus, the trade expansion/contraction cycle might cycle at 440 ± 80 years. This kind of coupled alternation of double negatives at doubled scales might account for J.S. Lee's long SPI-intense cycles at 880 ± 160 years and within them the 440 ± 80 year less-intense SPI cycles, along with trade region expansion/contraction cycles. Figures 8 and 11 show how Turchin cycles might be embedded inside the J.S. Lee and trade cycles

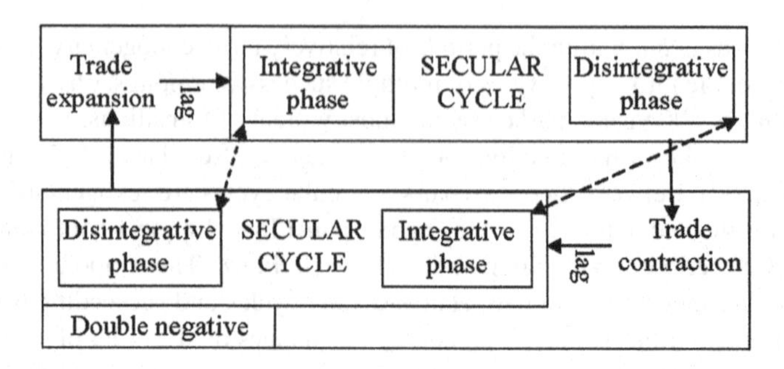

Fig. 10 Operation of a double negative as the key to 2:1 secular/city size phasing

that we believe have some affinity with our city system oscillations. Figure 10 gives an impressionistic model of how long cycles ca. 440 years might contain within ca. 220 year-cycles (given large margins of divergence, possibly contracting over time) in which the end of which very large super-regional trading systems alternately expand or contract. This observation is similar to, although at a larger scale, the cycles described by J.S. Lee [16] for China. Figure 10 expands (graphically) on the idea that: (1) a disintegrative phase in one of Turchin's [30, 32, 33] "secular cycles" might cause warfare and trade contraction; while (2) a succeeding disintegrative phase might undermine boundaries that are a barrier to trade and expand the inter-polity trading areas. As an explanation of a new cycle of an empire, Lee divides 800 year periods of Chinese history into two periods of 400: the first part is an early growth of an "empire" whereas the second part is a time of turbulence. His Socio-Political Instability data are shown in Fig. 11 (top) for China, and our coding (bottom) of this variable for Europe. While speculative, this kind of idea is not dissimilar to those of Modelski and Thompson [20] who find temporal "doublings" of systemic changes in which opposing tendencies occur at different time scales.

Further evidence for the coupling of historical urban system dynamics with structural demographic population dynamics [13, 14, 32] is evident in Fig. 12 just as in Turchin's historical dynamic models SPI leads population declines. The top graph in the figure shows that SPI is synchronously correlated with low β (reduced urban hierarchy) in city distributions for China. β recovers following peaks in SPI, just as population does. The bottom graph in Fig. 12 shows a 50-year lag between a high q/β ratio and rise in SPI, much as population growth relative to resources predicts rise in SPI.

Figure 4 has shown that q and β vary somewhat independently, correlated positively but only moderately. But which affects which over time if the two are

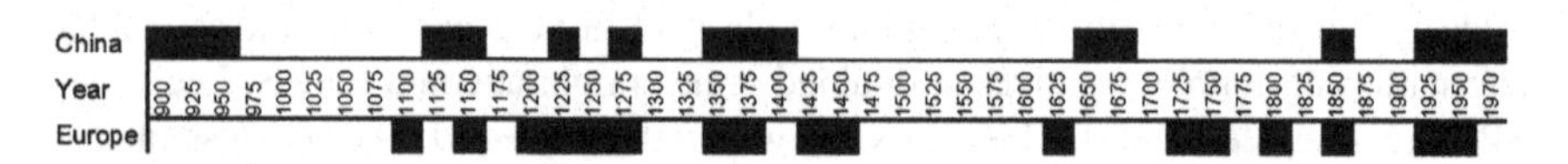

Fig. 11 Socio-political instability index (I)

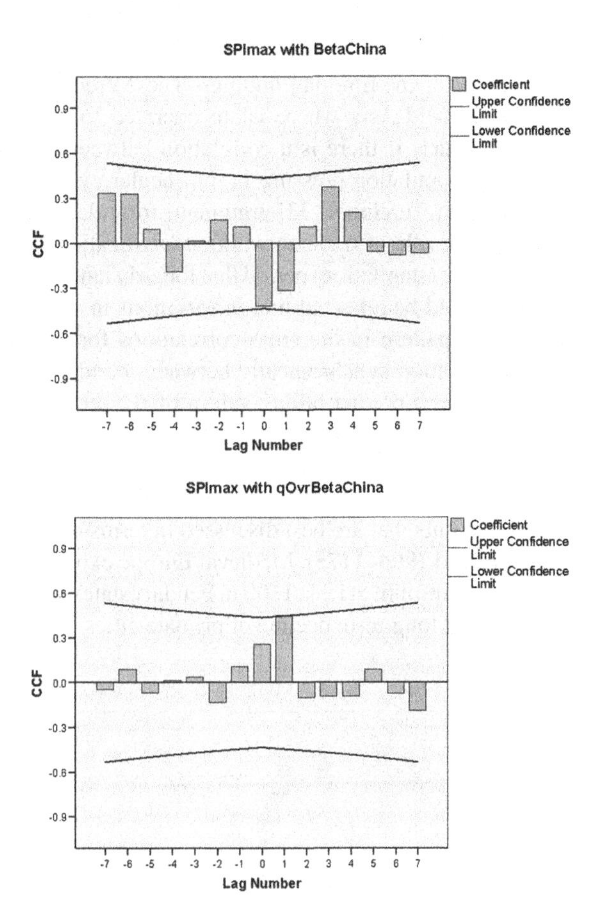

Fig. 12 Cross-correlations of q and β and sociopolitical instability (SPI). The *top graph* shows that SPI is synchronously correlated with low β (reduced urban hierarchy) in city distributions for China. β recovers following peaks in SPI, just as population does. The *bottom graph* shows a 50-year lag between a high q/β ratio and rise in SPI, much as population growth relative to resources predicts rise in SPI

synchronously somewhat independent? In time-lagged correlation: Does the shape (q) of the body of the city affect the tail (β) in subsequent periods, or the reverse? β might shape q if long distance trade has an effect on the larger cities engaged in international trade, but q might shape β if it is the waxing and waning of industries in the smaller cities that feed into the export products for the larger cities, as we often see in China and Europe.

The lagged cross-correlations for China and Europe—but not in Mid-Asia—show that, starting from the maximal correlation at lag 0, high q (e.g., over 1.5) predicts falls in Pareto β over time, reducing the slope of the power law tail below that of the Zipfian. This suggests that high q produces an urban system decline in β. This would contradict a hypothesis of long-distance trade as a driver of rise and fall in the larger cities. It would not contradict, however, the possibility that long-distance trade was directly beneficial to the smaller cities, with these effects feeding into the success of the larger cities with a time lag. For China and Europe, where successful long-distance trade was organized on the basis of the diffusion of effective credit mechanisms available to the smaller merchant cities, this seems a plausible explanation for the time-lag findings. These credit mechanisms were not so easily available in Mid-Asia where Islam operated to regulate interest rates to prevent excessive usury. If there is a correlation between long-distance trade and the rise and fall of population pressure in the secular cycles of agrarian empires, our data might support Turchin's [33] argument, formulated partly in response to our own studies of the role of trade networks in civilizational dynamics that it is during the high-pressure (stagflation) period that long-distance trade flourishes. If so, the impact of trade should be reflected first in variations in q, which vary more slowly than β. The overall pattern in the cross-correlations for the three regions together shows strong correlation synchronically between q and β at lag 0 ($p < 0.000001$) while high values of q predict falling values of β over time over three 50-year lags.[11] A dynamic including q and β, however, is not that simple.[12]

Figure 13 plots the fitted values of q and β, showing how changes in q lead β for both China and Europe. It shows that q leads β for Europe and China, with different inflection points that are best discussed in terms of historical periods of inflection:

PERIOD 1 (900–1175). Medieval Europe experienced an agricultural boom that would continue until at least 1100 in Feudal estates with towns having Zipfian distributions (q) but long-term decline of primate cities (β) continued. China after 900 AD

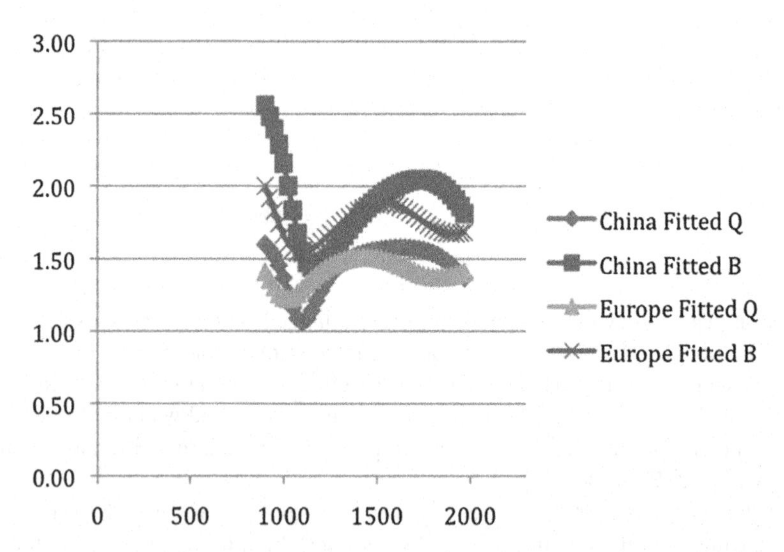

Fig. 13 Changes in q and β lead differently over time for China and Europe. These graphs show three large periods, the last two divided into two sub-periods, one of rise, then one of fall

[11] Variables q and β have the lowest cross-correlations for Mid-Asia, but detailed examination of the Mid-Asia time-lags shows a weak cyclical dynamic of Hi-q→Lo-β→Lo-q→Hi-β that holds to 1950.

[12] The dynamics is analyzed using an extrapolated Chandler [9] data such that the time lag is at 25 years.

is in decline after the fall of the Tang Dynasty, with q falling for 2.5 centuries. The growth of Kingdoms and bullion wealth exchanged for Silk Road luxuries supported the early Song period with growth coming from Silk Road trade in the western region. Overprinting money and overextending credit by the Dynasty maxed out with internal Socio-Political Conflict (peasant entrepreneurs vs. Song Dynasty enriched by their long-distance trade) and Mongol conquest of the Northern Sung. The new capital city of the Southern Song Dynasty created inflationary financial and currency collapse, and then recovers in a period of internal (1175–1250) and external (1238–1250) conflict. Europe's decline in β is preceded by declining β in China and lasts an extra century, β lagging q, i.e., merchants/peasants affecting wealthy urbanites.

PERIOD 2 (1175–1400). This is a turnaround period for the European medieval Renaissance and for China in terms of β turning upward towards the Zipfian. While q fluctuates in Europe with inter-city wars, q falls with conflicts and then rises above the Zipfian in China. This is brought on by the Yuan Mongol Dynasty (1271–1386) building of the Grand Canal, which was a catalyst of growth for the small/midsize cities along it and it fed in turn the growth of large cities, which continues to the next periods.

PERIOD 3 (1400–1450). This is a short period of time but one of a significant change in parameters, one of opposites for Europe and now Ming Dynasty China, the latter gaining super-Zipfian urban hierarchies for a century, retreating from global exploration in 1450. The Chinese mega projects (palaces, etc.) and building of big cities, especially Beijing, as well as the fortification of the Great Wall, used resources and materials from all over China, transported by the Grand Canal. This has much to do with q leading β, although here it is polity associated with β constructing q-development. European efforts after 1400, in contrast, beginning with the Portuguese Kingdom, engage them in maritime struggles against Asian domination that cost them sharply diminishing sub-Zipfian q and β levels. This period is followed by different trajectories for Europe and East Asia, with Europe opening Asian and New World conquests that in the long-term will raise q to Zipfian levels and unrelenting rises toward Zipfian large-city β.

PERIOD 4 (1450–1650+). For Europe this is a major turnaround period after 1450 with Maritime conquests that continue for centuries. In contrast, the Ming Chinese struggled to consolidate their hold on China due to political power struggle (eunuchs took over the government), economic crisis (sudden unavailability of the currency medium: silver), and natural disasters. In this period, China still retained above-Zipfian small-city q. However, q without long-distance trade no longer retains its effects in raising β, which declines up to 1644 when the Manchurian Qing Dynasty took over China.

PERIOD 5 (1650–1850). Chinese β rises up to the time of the British Opium Wars. And then drops again. From 1450–1850, fluctuations in β given dominance of the Qing capitals over Chinese in the smaller cities may be largely due to socio-political instability raised by various anti-Qing rebellions by the native Han.

PERIOD 5 (1850–). After 1850, according to Chandler and U.N. data (an adjustment between them handles differing distribution coefficients for the same year), China values of q have fallen and risen (subtracting the post-WW I year of 1925) as have those of β. For Europe q has fluctuated while those of β have consistently declined to 1.4, below Zipfian (2.0). There is not, in detail, a tendency for β to follow q.

Figure 13 showed general patterns of how small-city q tends to lead changes in city hierarchy β for both Europe and China. We now turn to the dynamics of these changes including the relation to sociopolitical violence, which has punctuated the discussion in our historical period commentaries.

Two-equation time-lagged regressions with constants fitted to each term behave similarly for China and Europe, 925–1970, as shown in Eqs. (3) and (4), adding a fitted constant C for the third term (β) and signs for the other three terms, except that a binarized version of the SPI index, I, designed to measure changes that would affect city populations, affects q without a time lag.[13] Sociopolitical instability I tends to have an immediate effect on the relation between smaller and larger cities that affects q but not β[14]:

$$\begin{aligned}&\beta(\mathrm{t}+1) \leftarrow -q(t) + q(t)\beta(t) + C\beta \\ &(\text{overall } \mathrm{R}^2 \sim 0.78, \text{China} \sim 0.74, \text{Europe} \sim 0.67)\end{aligned} \tag{3}$$

$$\begin{aligned}&q(\mathrm{t}+1) \leftarrow -\beta(t) + q(t)\beta(t) + Cq - I(\mathrm{t}+1) \\ &(\text{overall } \mathrm{R}^2 \sim 0.50, \text{China} \sim 0.43, \text{Europe} \sim 0.62)\end{aligned} \tag{4}$$

Without the effect of SPI (denoted by I), these two equations, unlike (1) and (2), would predict positive feedback between β and q that would result in either a convergent or a divergent time series.[15]

Figure 14 shows that the SPI index (I) acts as an external shock that creates, given the locations of SPI events, the Turchin oscillatory dynamics between q and β in the predicted time series. Without external shocks other than SPI (I) in equation (4), the model represents the likelihood that without external war or natural disasters like flooding and drought, the value of β would converge to a (Pareto I) Zipfian tendency with $\beta = 2$. Ran on a t-test, this oscillatory yields to around $q = 1.45 \pm 0.19$ and variations around $\beta = 1.78 \pm 0.35$.

This model adds support for the conflict events within Turchin's endogenous dynamic as the driver of q in the city dynamic, which in turn drives β (Fig. 13 and

[13] Socio-political instability index (SPI), I, is binary with a value of 1 to indicate instability and 0 otherwise. The rules to indicate instability are as follows: (1) A higher number of wars that are not isolated and affected the areas of conflicting interests; (2) natural disasters that significantly affect population (ex: the Black Death); (3) a significant mobility of population due to domestic socio-political unrest (ex: the movement of Song Dynasty's capital from north to south).

[14] There are two additional constants ($C\beta$ and Cq). All coefficients have $p < 0.021$ for both equations. The specific coefficients are −0.441, 0.363, 1.316, and −0.088 for Eq. (1) and −0.907, 0.573, and 1.580 for Eq. (2).

[15] A two-equation reciprocal time-lag model such as Eqs. (1) and (2) produces fluctuations if the signs of the right hand elements are opposite, but convergence or divergence if they are the same. This can be verified in difference equations using initial values that generate a full time series.

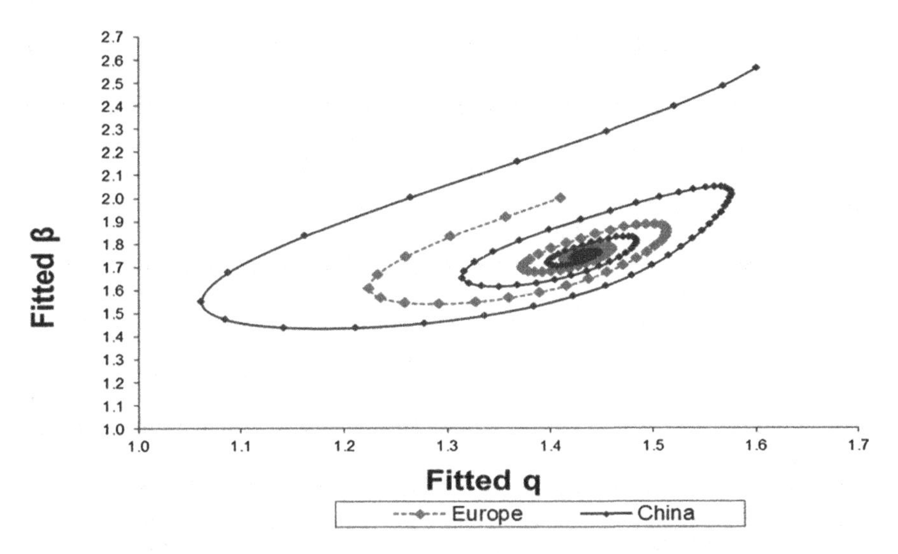

Fig. 14 Fitted time-series endogenous dynamic (inward spiral) of β and q when there is no exogenous shock caused by population pressure and/or instability. The *upper-right blue* and *red* nodes represent China $\beta(t)$ for 900 AD and Europe $\beta(t)$ for 1100 AD. The first *blue* inflection point at which q changes direction for China is at 1175, two periods after the inflection on the graph (1500 is among—six before—the closely packed points at the second Chinese inflection). The first *red* inflection for Europe is at 1200, one before the inflection on the graph (1500 is also among—six before—the closely packed points at the second European inflection). The second (Chinese and European) inflection occurs at year 1700 in both inward spirals, a point at which upward swings of β recur in synchrony for the second time

Eq. 3). The three main factors that make for instability are external invasion (war), natural disaster, and domestic socio-political unrest. Their periods of collapse are affected synchronically with generational time lag by population pressure P and SPI (I) consistent with the dynamics of Turchin's model of agrarian empires. In Figs. 13 and 14, China has a greater oscillation than Europe because China had rapid growth in 1300–1875 compared to Europe, which had slumps in 1400–1500, as explained by the historical periods mentioned above.

Figure 14 shows an overall near-Zipfian (1.4–1.5) inward-spiral tendency for q, shaping the body of city-hierarchy, and less than Zipfian (2.0) for β, (1.7), capturing the thinner than expected primate cities slope in the tail. These facts accentuate our arguments that these might show the less than Zipfian optimum growth structure of city-hierarchy, especially for top echelon of primate cities, before they are swept away in a short period by interregional competition due to socio-political instability.

PERIOD 1 (900–1175). The year 900 is the upper left terminus of the Fig. 14 curve, governed by Eq. (3) where the product $q(\mathrm{t})\beta(\mathrm{t})$ is large in year 900 for China (blue line), and as q drops β also drops, until both reach maximum sub-Zipfian levels of collapse in q at the first bend (lower left) of the figure.

PERIOD 2 (1175–1400). Now Eq. (4) takes over, with the Fig. 14 climb in q values starting an upward climb, also halfway toward the β Zipfian (2.0) by 1400, close to the cusp of the next sharp change in the spiral. This rise continues over a shorter period (1170–1325) in China but a longer period in Europe (1250–1400, β leading q, again with the C_q and $q(\mathrm{t})\beta(\mathrm{t})$ outweighing the contrarian effect of $-\beta(\mathrm{t})$ on $q(\mathrm{t}+1)$.

PERIOD(S) 3–5 (1400–1950). The upward curve of Eq. (4) now rounds the spiral at its second sharp bend, now with city sizes q and β falling in a globalizing context of recurrent conflicts that continues into and a century beyond the present. That would predict further sub-Zipfian distributions in city-size hierarchies.

Equations (3) and (4), that is, in Fig. 14, do not recover the finer-grained periodization that we identified as periods 3–5 from Figs. 5 and 6, and seen as finer modulations of Fig. 13, where there are curvatures that help to suggest that changes in q tend to precede changes in β that are in the same direction.

Note that over these three longer periods, Song China starts out with entrepreneurship led by peasant/merchants' actually lowering q (Silk trade independent of the Dynastic polity) and as q and β rise, cities become top heavy with Dynasties and Empires. Europe starts out in this period with Feudal fiefdoms, Lords and Monarchs, and becomes dominated by merchants, especially evident in the Northern Italian Renaissance and later in the 1680 "Glorious Revolution" of England, which toppled the landed aristocracy in Periods 3 and 4.

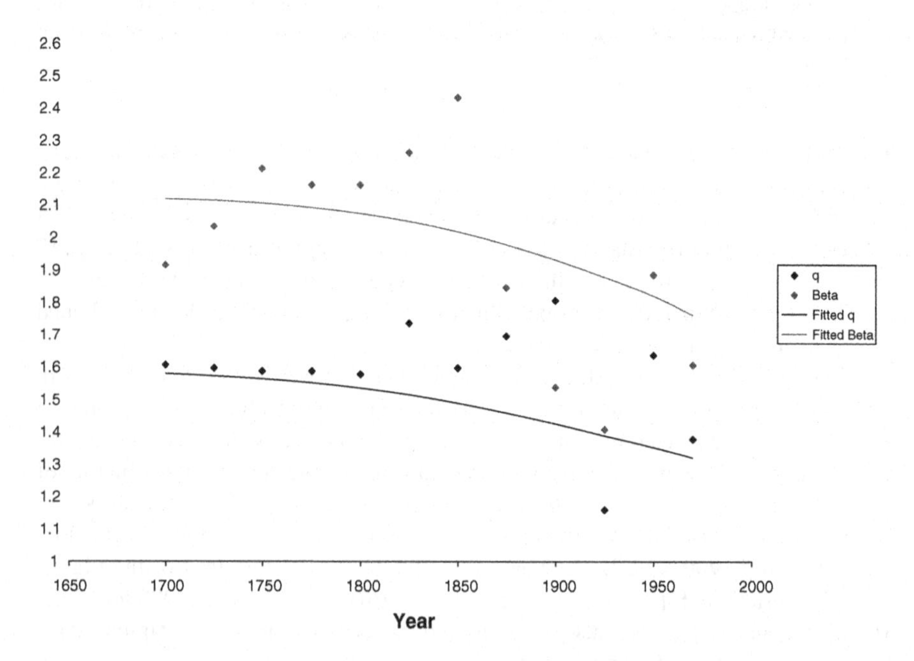

Fig. 15 China time-series Fitted by Eqs. (3) and (4) as in Fig. 14 but starting at 1700

Figure 14 can mislead the reader into thinking that convergence in values of q and β increases with time. Figure 15, which starts the fitting of Eqs. (3) and (4) at 1700, is designed to correct that impression. The equations fit part of a smooth inward-spiral graph but the actual data points here for Chinese q and β do not converge as do their fitted values. These data points also finish at the upper right top of the second turn in the blue curve (q) in Fig. 14, whereas the inward spiral continues in the future for hundreds more hypothetical years.

9 Methods and Results

Our comments on the cross-regional effects of Europe on China in various of our Time Period discussions can be balanced by our previous work [36, Figs. 9.7–9.9 and 9.12] using the autocorrelation function (ACF) to test whether the effects of time-series values of a variable in one time period are correlated in value successive time lags (each lag in this case adding 50 years). These tests give upper and lower confidence limits at 95 % for a two-tailed significance test ($p < 0.05$). The ACF of β compared to lags of q shows a short-term continuity effect (1 lag of 50 years). Results include for China: the SPI lowers β significantly and the SPI/β ratio predicts higher SPI (sociopolitical violence) after a 50 year time lag [36, p. 216]; q in China significantly affects β in Europe with a 100 year time lag [36, p. 210]; Silk Road trading intensity significantly affects European q at 50 and 100 year lags [36, p. 211]. While not discussed here, we also found that MidAsia q had a significant effect on q for China and for Europe.

Our results, measurements, and mathematical models compare favorably with of the structural demographic ("secular cycles") studies of Goldstone [13], Nefedov [21], and Turchin [30, 32]. Our cycle of city-size oscillations might be twice as long as Turchin's secular cycles. One possible explanation is that our city-system fluctuations are exogenously affected by the conflict phase of his structural demographic cycles, but these conflicts might or might not pass the threshold to have such effects. From both perspectives, however, sociopolitical instability is not smoothly cyclical but episodic. Rebellions, insurrections, and all sorts of protest are events that mobilize people in a given time and generation, and that impacts that, when repeated frequently, have massive effects. We see this in long-term correlations with SPI, such as internecine wars in China.

We have been able to discern some of the effects of trade fluctuations (if not trade network structure) in these models. Some of the patterns we see in our data that concern globalizing modernization are consistent with prior knowledge and others are startling. To be expected are the developmental trends of scale—larger global cities, larger total urban population, and larger total population. What is startling is that there are some long-wave oscillations in q that are very long. Hopefully, a long-term trend and contemporary structure of Zipfian city distributions is an indicator of stability, but even the twentieth century data indicate that instabilities are still very much present and thus likely to rest on historical contingencies (somewhat like the occurrence of a next earthquake larger than any seen in x years prior), and very much open to the effects of warfare and internal conflicts that are likely to be affected by population growth, and as opposed to stabilization of trade benefiting per-capita-resources ratios.

10 Conclusion

Our results allow us to consider the edge-of-chaos metaphor of complexity with respect to q as a first approximation for modeling dynamical instabilities of city-size distributions historically which are nonetheless metastable and possibly converging to greater stability. It is a truism to say that complexity, life, history, and complex systems generally stand somewhere between rigidity at one pole, which might be exemplified by $q > 1.8$, and on the other, an exponential random distribution ($q \approx 1$) of city sizes, or the unpredictability of chaos ($0 < q \ll 1$).[16] But we do not see support for equilibrium on the "edge of chaos" in these data. Instead, we see the making of an empirically-based dynamical model with variability around the Zipfian based on the interactions of the two parameters of the q-exponential with variables such as trade and war. The historical q-periods of China and other regions tend to cluster, somewhat like "edging on chaos," near an average of $q \approx 1.5$, but they do so in terms of oscillations, not far from equilibrium, and not a stable equilibrium. This is in tandem with oscillations around $\beta \approx 1.78$ (± 0.35) showing the less than Zipfian $\beta = 2$ optimum growth structure for the top echelons of primate cities. From that average they may fall into a more decentralized state nearer chaos (with exceptionally low q), in which the power-law tails are minimal (with larger cities crushed in size by internecine wars or smaller cities are very common relative to hubs), or rise into regimes affected by massive external drains on the economy or political policies that seem to put q into abnormally rigid states of high q. The directions of change in q are largely predictable as a function of the current-state variables (such as population/resource ratios and sociopolitical violence) in the historical dynamics models up to, but not yet including, the contemporary period. How to derive predictions for the contemporary era is not yet evident given the new configurations of industrial societies, but it is very probable that such predictions as do emerge for the present will contain processes operative in the past.

Perhaps the most important of our results is the finding that q leads β in city development. This is consonant with Jacobs [15] who stressed the reliance of cities on their surrounding economies and networks. Economists and physicists who theorize about Zipfian city distributions neglect the importance of the lower part of city-size distributions that is captured by q-exponentials and may thereby fail to be aware or see in their quantitative work the effects of smaller cities in urban and economic dynamics.

[16] Technically, the mathematics assigned to chaos is a deterministic departure from randomness in which a dynamic trajectory never settles down into equilibrium, and small differences in initial conditions lead to divergent trajectories. The link between empirical history and "edge of chaos" is typically done by simulation.

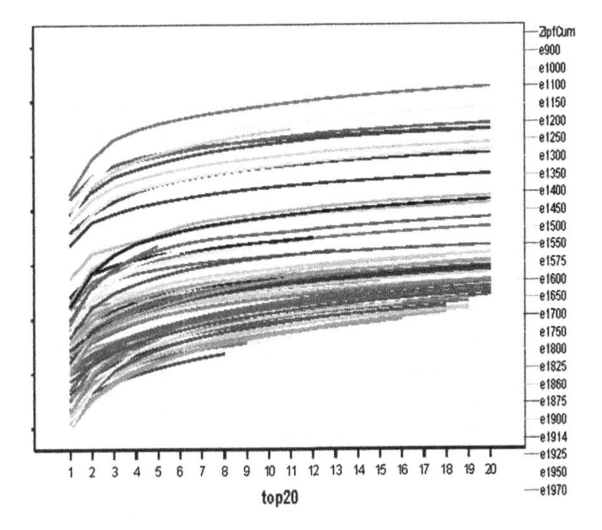

Fig. 16 The Chandler Rank-Size City Data (semilog) for Eurasia

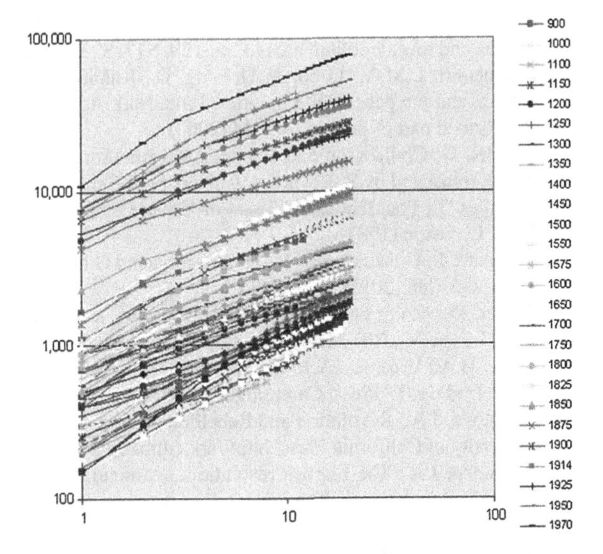

Fig. 17 The Chandler Rank-Size City Data (log-log) for Eurasia

11 Appendix A: Example of the European City-Size Distribution

Figures 16 and 17 show the raw European cumulative city-size distributions truncated to the top 20 cities for each of Chandler's historical periods. The x-axis gives city rank. The y-axis in Fig. 16 is the natural log e of city size, unlabeled. Figure 17 shows the same data in a log10-log10 plot, where city rank is also logged and the y axis should be multiplied by 100. If the latter plot showed straight lines, the distributions would be Pareto I power-laws. Cutoffs can be seen for each of the lines, an indication that the Pareto II q-exponential is more relevant for these distributions.

Minimum top-ranked city-sizes recorded by Chandler [9] are as small as 11,000 persons, but not lower, so that cities of sizes down to 5,000 are not included. Although the q-exponential has two parameters, a power-law fit to these data would also require a second parameter for each distribution, the size at which the distributions of higher sizes is power-law.

References

1. Bairoch, P.: Cities and Economic Development: From the Dawn of History to the Present. University of Chicago Press, Chicago (1988)
2. Batty, M.: Rank clocks. Nature (Letters) **444**, 592–596 (2006)
3. Batty, M.: The size, scale, and shape of cities. Science **319**(5864), 769–771 (2008)
4. Batty, M., Longley, P.A.: Fractal Cities: A Geometry of Form and Function. Academic Press, London (1994)
5. Bercher, J.-F., Vignat, C.: A new look at q-exponential distributions via excess statistics. Physica A **387**(22), 5422–5432 (2008)
6. Berry, B.J.L., Okulicz-Kozaryn, A.: The city size distribution debate: resolution for US urban regions and megalopolitan areas. Cities **29**, S17–S23 (2012)
7. Bettencourt, L.M.A., Lobo, J., Helbing, D., Kühnert, C., West, G.B.: Growth, innovation, scaling, and the pace of life in cities. Proc. Natl. Acad. Sci. U S A **104**, 7301. This research also formed part of project ISCOM (2007)
8. Braudel, F.: Civilization and Capitalism, 15th-18th Century, vol. 3. The Perspective of the World, translated by S. Reynolds. University of California Press, Berkeley, CA (1992)
9. Chandler, T.: Four Thousand Years of Urban Growth: An Historical Census. Edwin Mellon Press, Lewiston (1987)
10. Clemente, J., Gonzalez-Val, R., et al.: Zipf's and Gibrat's laws for migrations. Ann. Reg. Sci. **47**(1), 235–248 (2011)
11. Farmer, D.: Using behavioral statistical physics to understand supply and demand. Invited Paper for the MAR07 Meeting of the American Physical Society (2007)
12. Fujita, M.A., Venables, A.J., Krugman, P.: The Spatial Economy: Cities. Regions and International Trade. MIT Press, Cambridge (1999)
13. Goldstone, J.A.: Revolution and Rebellion in the Early Modern World, eScholarship edition. University of California Press. http://ark.cdlib.org/ark:/13030/ft9k4009kq/ (1991)
14. Goldstone, J.A.: The English revolution: a structural-demographic approach. In: Goldstone, J.A. (ed.) Revolutions—Theoretical, Comparative, and Historical Studies, 3rd edn. University of California Press, Berkeley (2003)
15. Jacobs, J.: The Death and Life of Great American Cities. Random House, New York (1961)
16. Lee, J.S.: The periodic recurrence of internecine wars in China, The China J. (March–April), 111–163 (1931)
17. Malacarne, L.C., Mendes, R.S., Lenzi, E.K.: q-exponential distribution in urban agglomeration. Phys. Rev. E **65** (1 Pt 2 017106), 1–3 (2002)
18. Modelski, G.: World Cities –3000 to 2000. Faros 2000, Washington, DC (2000)
19. Modelski, G.: Evolution of the world economy. Paper prepared for the Session on Social Dynamics and the Encyclopedia of Human Ecology: A Kenneth Boulding Retrospective, Boston, MA, 7–9 January 2000. http://faculty.washington.edu/modelski/Evoweconomy.html (2000)
20. Modelski, G., Thompson, W.R.: Leading Sectors and World Powers: The Coevolution of Global Politics and Economics. Univ. of South Carolina Press, Columbia (1996)
21. Nefedov, S.: The method of demographic cycles in a study of socioeconomic history of preindustrial society. Ph.D. Dissertation, Ekaterinburg University, Ekaterinburg, Russia (1999)

22. Newman, M.E.J.: Power laws, Pareto distributions and Zipf's law. Contemp. Phys. **46**, 323–351 (2005)
23. Pareto, V.: La courbe des revenus. Le Monde economique **6**, 99–100 (1896)
24. Pasciuti, D.: Estimating population sizes of largest cities. Workshop Paper for Measuring and Modeling Cycles of State Formation, Decline, and Upward Sweeps Since the Bronze Age, San Diego, March 2006
25. Picoli Jr., S., Mendes, R.S., Malacarne, L.C.: q-exponential, Weibull, and q-Weibull distributions: an empirical analysis. Physica A **324**, 678–688 (2003)
26. Reggiani, A., Nijkamp, P.: Did Zipf anticipate socio-economic spatial networks? Quaderni DSE Working Paper No. 816 (2012)
27. Shalizi, C.: Maximum Likelihood Estimation for q-Exponential (Tsallis) Distributions (Code at http://www.cscs.umich.edu/crshalizi/research/tsallis-MLE/), Carnegie Mellon University, Statistics Department, math/0701854 (2007)
28. Tsallis, C.: Possible generalization of Boltzmann-Gibbs statistics. J. Stat. Phys. **52**, 479–487 (1988)
29. Tsallis, C.: Historical Background and Physical Motivations. In: Gell-Mann, M., Tsallis, C. Nonextensive Entropy: Interdisciplinary Applications. Oxford: Oxford University Press. (2004)
30. Turchin, P.: Historical Dynamics: Why States Rise and Fall. Cambridge University Press, Cambridge (2003)
31. Turchin, P.: Meta-ethnic Frontiers: Conflict Boundaries in Agrarian Empires, 0–2000 CE. Powerpoint slides, Santa Fe Institute Working Group on Analyzing Complex Macrosocial Systems (2004)
32. Turchin, P.: Dynamical feedbacks between population growth and sociopolitical instability in agrarian states. Struct. Dyn. **1**(1), 49–69 (2005)
33. Turchin, P.: Modeling periodic waves of integration in the Afroeurasian world-system. In: Modelski, G., Devezas, T., Thompson, W.R. (eds.) Globalization as Evolutionary Process: Modeling Global Change. Routledge, London (2008)
34. White, D.R.: Innovation in the context of networks, hierarchies, and cohesion. In: Lane, D., Pumain, D., van der Leeuw, S., West, G. (eds.) Complexity Perspectives in Innovation and Social Change, Methodos Series, Chap. 5, pp. 153–193. Springer, Berlin (2009)
35. White, D.R.: Networks and globalization policies. In: Vedres, B., Scotti, M. (eds.) Networks in Social Policy Problems. Distributions in Empirical Data. Cambridge University Press, Cambridge (2012)
36. White, D.R., Tambayong, L., Kejžar, N.: Oscillatory dynamics of city-size distributions in world historical systems. In: Modelski, G., Devezas, T., Thompson, W.R. (eds.) Globalization as Evolutionary Process: Modeling Global Change, Chap. 9, pp. 190–225. Routledge, London (2008)
37. White, D.R., Kejžar, N., Tsallis, C., Rozenblat, C.: Generative historical model of city-size hierarchies: 430 BCE–2005. Project ISCOM Working Paper (ms.). Institute of Mathematical Behavioral Sciences, University of California, Irvine, CA. (2005)
38. Zipf, G.K.: Human Behavior and the Principle of Least Effort. Addison Wesley, Cambridge (1949)

Chapter 11
Preliminary Analytical Considerations in Designing a Terrorism and Extremism Online Network Extractor

Martin Bouchard, Kila Joffres and Richard Frank

Abstract It is now widely understood that extremists use the Internet in attempts to accomplish many of their objectives. In this chapter we present a web-crawler called the Terrorism and Extremism Network Extractor (TENE), designed to gather information about extremist activities on the Internet. In particular, this chapter will focus on how TENE may help differentiate terrorist websites from anti-terrorist websites by analyzing the context around the use of predetermined keywords found within the text of the webpage. We illustrate our strategy through a content analysis of four types of web-sites. One is a popular white supremacist website, another is a jihadist website, the third one is a terrorism-related news website, and the last one is an official counterterrorist website. To explore differences between these websites, the presence of, and context around 33 keywords was examined on both websites. It was found that certain words appear more often on one type of website than the other, and this may potentially serve as a good method for differentiating between terrorist websites and ones that simply refer to terrorist activities. For example, words such as "terrorist," "security," "mission," "intelligence," and "report," all appeared with much greater frequency on the counterterrorist website than the white supremacist or the jihadist websites. In addition, the white supremacist and the jihadist websites used words such as "destroy," "kill," and "attack" in a specific context: not to describe their activities or their members, but to portray themselves as victims. The future developments of TENE are discussed.

M. Bouchard (✉) · K. Joffres · R. Frank
School of Criminology, International CyberCrime Research Centre, Simon Fraser University, Burnaby, BCV5A 1S6, Canada
e-mail: mbouchard@sfu.ca

K. Joffres
e-mail: kja4@sfu.ca

R. Frank
e-mail: rfrank@sfu.ca

V. K. Mago and V. Dabbaghian (eds.), *Computational Models of Complex Systems*, Intelligent Systems Reference Library 53, DOI: 10.1007/978-3-319-01285-8_11,
© Springer International Publishing Switzerland 2014

1 Introduction

Much like any other groups and organizations, extremist and terrorist groups can be found on the Internet, including many who have their official website [21]. Conway [4] has suggested that extremists use the Internet in five general ways: information recruitment, networking, information provision, financing, and recruitment (see also [16]). The Internet's appeal follows from its ability to provide a broad reach, to provide low costs, to be timely and efficient, and to provide some degree of security and anonymity [8]. Tsfati and Weimann [19] further emphasize that the Internet is extremely well suited to terrorists for the purposes of communication as it is decentralized, uncensored, largely free of control or restrictions, and allows for worldwide access for current members or potential recruits

Despite the identification of the Internet as a tool for terrorist groups, limited empirical research has been conducted into the nature of the terrorism-related content online. However, researchers involved with the Dark Web Project have started to build knowledge on the content and structure of websites hosting terrorism-related content [1]. For example, Zhou et al. [24] proposed a semiautomated methodology (combining the efficiency of automatic data collection and the accuracy of manual collection) for identifying, classifying, and organizing extremist Website data. The Dark Web Project has since spawned several research studies on the nature of terrorist use of the Internet and their online networks [1]. These methodologies allow for the analysis of extremist Website content, which can further inform counterterrorism research and policy (e.g., in terms of disrupting online networks and intelligence-gathering on terrorist operations).

Information on terrorist communications, ideologies, activities, relationships, and so on, can be produced by studying extremist web-sites. As such, the main objective of this chapter is to lay the foundation for the development of the Terrorism and Extremism Network Extractor (TENE). In its fully developed form, TENE will be a custom-written computer-program that automatically browses the World Wide Web for terrorism and extremism content, collecting information about the pages it visits.

The foundation of this approach is based on a combination of the ground breaking work associated with the Dark Web Project [1] and our previous efforts to study online child exploitation websites [6, 10, 22]. For instance, the Dark Web project has spawned various programs designed to examine particular types of online extremist content (e.g., forums), to visualize and determine the topology of networks of extremist webpages, and to study certain aspect of the content on extremist websites (e.g., technical sophistication, media richness, and web interactivity) [1]. However, an "intelligent" version of TENE that would browse the Web to automatically identify extremism content has yet to be created. Methods to allow TENE the capability to make decisions automatically can be done, for example, by borrowing techniques from the domains of datamining [9, 11] or machine learning [15, 18]. Features of webpages to use for analysis could include keywords, page structure and the location of that page within the Internet. While analyzing the frequency and occurrences of certain keywords on a website may assist in suggesting something about the nature of the content, it is unlikely to be enough for accurately labeling a website as containing and spreading extremist views. A news website, for example, may contain the same words with similar frequencies. The same may potentially be said of a governmental website focused on issues associated with terrorism and counter-terrorism. Unfortunately, the type and content of websites containing material that can be qualified as "extremism" has yet to be fully understood (the aphorism that one man's terrorist is another man's freedom fighter has yet to find a suitable empirical solution for such purposes). Thus, before undertaking the process of making TENE more intelligent, additional baseline data is needed about the structure and content of single websites in order to establish the ground rules necessary to create a valid web-crawler program.

The current study contributes to this end goal. In this chapter, we use a preliminary version of the web-crawler designed to collect the entire information contained on a single website. In particular, this chapter will focus on how TENE may help differentiate terrorist/extremist websites from other types of related websites by analyzing the context around the use of predetermined keywords found within the HTML of the webpage. We illustrate our strategy through a content analysis of four types of websites. One is a popular white supremacist website, another is a jihadist website, the third one is a terrorism-related news website, and the last one is an official counterterrorist website.

2 Methodology

TENE is a web-crawler that emerges from previous work on extracting online child pornography networks (see [6, 10, 22]). TENE operates by starting the crawling process at user-specified webpages, retrieving the pages from the Internet, analyzing them, and recursively following the links out of the pages. For the purpose of this chapter, the web-crawler starts at a page that covers material broadly associated with extremism or terrorism. Such a webpage can be found by the user, given to the web-crawler by the police, or obtained from terrorism-related literature. The starting website is then retrieved for the crawler, but there is no need to display the content in a web-browser and hence only the HTML (Hypertext Markup Language) of the webpage is retrieved. Certain statistics about the content of webpages are recorded, such as the frequency of user-specified keywords and count of images or videos. In its mature form, TENE will also follow the links found on a webpage if these links point to a webpage that contains extremism or terrorism material. These links will be subsequently explored recursively until certain criteria are met.

As the Internet is extremely large and a crawler would most likely never stop crawling, three conditional limits can be implemented into the web-crawler. These conditions help keep the crawling process under control and the network content-relevant. First, to keep the network extraction time bounded, a limit can be put on the number of pages retrieved (in our previous work on child pornography, that limit was 250,000). Second, the network size may be fixed at a specific number of websites (for example, 500). The webpages are retrieved in such a way that each website is sampled equally, or as equally as possible. Finally, in order to provide some boundaries for the crawl and guide the network extraction process to a relevant network, a set of keywords needs to be defined. For the crawler to include a given webpage in the analysis, the page has to contain a user-defined number of unique keywords.

The end result of the crawling process is knowledge about a set of web-servers, including the webpages contained within them, and the links between the webpages. These results are then aggregated up to the server level, with the resulting network summarizing the content on each of the servers, count of keywords, videos, and images, and the links between each of the servers. This essentially creates a map of a terrorism network from the Internet. Note that the version of TENE used for the purpose of this chapter remains within the realm of the initial user-specified website from which it starts. This work will eventually lead to the establishment of rules allowing for automatic identification of a terrorism/extremism-related website from another.

2.1 Keywords

As previously mentioned, to keep the websites crawled in TENE topic-relevant, keywords are used as an inclusion criterion. Previously, with the child exploitation application, specific keywords and their frequencies were used to establish it as a child exploitation website. The keywords were derived from manual analysis of the textual content on child exploitation pages, as well as through contacts with law enforcement. A new set of keywords is derived in this chapter from the terrorism and extremism domain, and includes words that extremist groups are known to include on webpages, such as bomb, recruit, attack, or target (full list can be found in Table 2). A rate of use per webpage will be calculated for each keyword, on each website, in order to detect whether a type of website is more likely to use a word compared to another. The root word method is used, so that "bomb" and "bombing," for example, is meant to be the same word.

Counting the number of instances a keyword was used on a webpage was sufficient in the child exploitation domain because the analysis showed that the keywords accurately indicated the presence of child exploitation, and were very poor at detecting the content of websites dedicated to countering child exploitation. This is expected to be more of a challenge in the terrorism domain, where words such as bomb or attack can be used, for example, by law enforcement-related websites focused on counter-terrorism. As such, TENE has been extended to not only count the number of instances of a keyword, but also capture the circumstances within which the keyword appears. That is, TENE is designed to be content-aware and at present, extracts a minimum of 200 characters before and after a given keyword. This number of characters was determined to be adequate for ascertaining the context of the keywords within their sentences, and with regard to immediately preceding and following sentences. This will allow for an analysis of each keyword in its original context to better understand when and how it is used. It should be noted that while

TENE is programmed to extract the context, the analysis of this context is performed manually. This step is expected to be replaced by some text-classification method(s) in TENE's final form.

2.2 *Websites Selected for Analysis*

Four different websites were explored using TENE, including a jihadist website, a white supremacist website, a news website, and a counterterrorist website. The jihadist and the white supremacist websites were obtained from research articles emerging from the Dark Web Project, and were labeled as "extremist" or "terrorist" websites by the researchers (see [2, 23]). The counterterrorist website was an official American government website while the news website was automatically identified by the web-crawler in a test run as having many of the keywords input into the program. Table 1 provides a brief description of each website, the number of pages crawled, and the website hosts. The jihadist website was quite small (nine pages in total), with the main page containing most of the information, supplemented by material from the other eight webpages within the website. The largest website (the news website) contained 164 pages of largely article archives.

3 Results

TENE crawled the webpages within the four aforementioned websites and recorded the number of keywords during this process. The keywords appearing on these websites and their context was explored in order to determine (a) whether these websites can be differentiated by the keywords that appear on them and (b) whether certain keywords are used in a similar or different manner across websites. Overall, our results show that certain keywords appear to be associated with specific websites. In addition, different websites use the same words in different context and occasionally use different words in the same context.

Table 2 shows the number of keywords mentioned per page; it can be seen that each website has keywords that tend to be specific to it. That is, certain keywords appear more frequently for each website compared to others. For instance, words such as "Allah", "attack", "Islam", "infidel", and "Jihad" occurred more often in the jihadist website than in any other website; these words ranged from 0.13 ("infidel") to 7.38 ("Islam") mentions per page. In contrast, the white supremacist website employed the words "Jew" (13.38 mentions per page) and "white" (20.98 mentions per page) considerably more often than other websites. The counterterrorist website also used specific words at a greater frequency, including "counterterrorism", "intelligence", "terrorist", and "security", and "combat". Conversely, words such as "dead", "kill", "foreign", and "Obama", appeared more commonly on the news website. This suggests that particular keywords may be used to distinguish between types of websites.

Table 1 A Description of the four websites selected for analysis

Website	No. of pages	Host	Category	Description
http://jorgevinhedo.sites.uol.com.br	9	The Lashkar-e Tayyiba, a prominent militant jihadist organizations in South Asia	Jihadist website [2, 23]	Posts information about Islam, attacks against Muslims, and includes links to Islamic newsletters and other relevant sources of information
http://www.natall.com	47	The National Alliance, a white supremacist and white nationalist political organization	White supremacist website [2, 23]	Includes online publications, broadcasts, a forum, and links to similar websites
http://www.counterpunch.org	164	Alexander Cockburn and Jeffrey St. Clair, editors of the website	News website	An American website that posts political news articles, updated every weekday
http://www.nctc.gov	15	The U.S. government. NCTC integrates and analyzes intelligence pertaining to terrorism, keeps a knowledge bank on terrorism information, and provides support to counterterrorism activities	Counterterrorist website	Includes information about NCTC, its mission, its goals, its products, and its activities. It also provides information on key partners, its director, and a page for children

Table 2 Number of keywords per page for each website

Keyword	Jihadist	White supremacist	News	Counterterrorist
Allah	3.63	0.04	1.07	0
Attack	6.38	0.89	0.87	0.6
Black	0.86	3.11	4.79	0
Bomb	0	0.28	0.8	0.53
Combat	0	0.06	0.1	0.47
Counterterrorism	0	0	0.02	4.33
Dead	0	0.15	1.60	0.07
Destroy	0.38	0.57	0.25	0
East	0	1.45	1.24	0
Fight	0	0.68	0.55	0.07
Foreign	0	0.19	0.59	0
Free Speech	0.5	0.04	0.40	0
Infidel	0.13	0.04	0	0
Intelligence	0	0.15	0.32	7.67
Islam	7.38	0.09	0.55	0
Jew	3.5	13.38	2.37	0
Jihad	2.25	0	0.06	0
Join	0.25	0.70	0.29	0.53
Kill	0.13	0.38	1.39	0.07
Live	0.25	1.55	0.91	0.07
Member	0.13	2.89	1.32	1.27
Mission	0	0.13	0.69	1.6
Obama	0	0.28	5.62	0.13
Race	0.38	19.28	11.12	0.67
Report	1	0.77	2.40	5.8
Security	0	0.17	1.04	4.07
Struggle	0.13	0.45	0.21	0
Terrorist	3.13	0.36	0.6	10.93
Train	0.25	0	0.39	0
Victim	0	0.15	0.26	0
Violence	0.63	0.38	0.55	0.13
West	1.5	1.72	1.99	0
White	0	20.98	0.70	5.27

This is represented more clearly in Table 3, where individual web-sites and combinations of websites are associated with particular keywords. This table was constructed by comparing the rates of words appearing for each website; if a word for a particular website occurred at a rate of less than one-third the rate of the highest incidence of that word, then no association between that word and the website was marked. This allowed us to classify words that were solely used by one of the four types of websites analyzed, as well as words that were used by various combinations of two or three web-sites. For instance, both the jihadist and the news website used the words "free speech" and "train" more frequently than other web-sites. Similarly,

Table 3 Keywords most strongly associated with specific websites

Website(s)	J	WS	N	C	J and N
Keywords	Allah	Jew	Dead	Combat	Free Speech
	Attack	White	Foreign	Counterterrorism	Train
	Infidel		Kill	Intelligence	
	Islam		Obama	Security	
	Jihad			Terrorist	
	WS & S	N & C	WS, N and C	J, WS and N	J, WS, N and C
	Black	Mission	Bomb	Destroy	Join
	East	Report	Member	Violence	
	Fight			West	
	Live				
	Race				
	Struggle				
	Victim				

J Jihadist; *WS* White supremacist; *N* News; *C* Counterterrorism

the white supremacist and the news website shared several words, including "black," "East," "fight," "live," "race," "struggle," and "victim." The news and counterterrorist website employed the terms "mission" and "report" more frequently than other websites. In some instances, all but one website used a particular word at similar rates. For example, the white supremacist website, the news website, and the counterterrorist website shared the words "bomb" and "member." The jihadist website, the white supremacist websites, and the news websites each used "destroy," "violence," and "west" at comparable rates. Note that the word "join" was similarly popular across all websites.

The type of keywords that appeared more frequently for each web-site followed intuitively from the nature of the website. For a jihadist website, words such as "Allah" and "Jihad" are more relevant to the subject matter of the website. While terms such as "white" and "Jew" may also be used in a jihadist website, these words are more likely to appear in the many discussions of "race" in the white supremacist website. However, both websites also use certain words at similar rates, including "destroy," "violence," and "West", and as will be seen, the websites tend to use such words in a similar context. The content of counterterrorist websites is very different, and the associated keywords reflect its focus on terrorism incidents and intelligence reports. For a news website, articles on deaths and politics are likely to be an important part of the website, and as such, are more popular words. In addition, both counterterrorist and news websites shared the use of "mission" and "report," which further differentiate their content from the more religious and racial focus of the jihadist and white supremacist websites. In these ways, the type of website tends to be represented in specific words that appear with greater frequency on the particular website compared to others.

4 Contextual Analysis

An initial crawl of the different websites also allowed for the examination of how certain keywords are used within each website. For example, certain words were used by the jihadist and white supremacist websites with similar frequency, but in varying contexts. Words such as "West," which were used with similar frequency (1.5 times per page for the jihadist extremist website and 1.72 times for the white supremacist website), took on very different meanings between the websites. In the jihadist website, "West" was used when discussing:

- "justification for Jihad against US and its terrorist western allies,"
- "anti-Islam sentiments all over the West," and
- "any invader in Afghanistan or Pakistan - be it West, India or Israel - would have to face the collective strength of the Muslims."

In essence, anti-West sentiments were prominent within the website. In contrast, the white supremacist website took pride in the West, discussing concerns about "alien groups" taking over "both in terms of culture and [white] genetic future," expressing concerns over "the decline of the West generally" with its growing multiculturalism, and posting assurances that the "National Alliance will lead [white] people to a secure homeland here in the Western hemisphere."

Other words were also used commonly between the various web-sites, but in different contexts. For instance, both the white supremacist and the news website used the word "black" more often than other websites (3.11 times per page for the white supremacist web-site and 4.79 times per page for the news website compared to 0.86 times per page for the jihadist website and 0 times for the counter-terrorist website). However, the news website used the word as a colour or adjective. In contrast, the white supremacist website largely employed the word to refer to African Americans, doing so in a typically disparaging manner (e.g., by discouraging interracial marriages, criticizing sympathetic portrayals of "Blacks" in the media and images of "Whites" and "Blacks" together, discussing the "problem" of "Black crime," and arguing that "Black History month is destroying the past").

While the same words were sometimes used differently between websites, other words were used to express the same point. For instance, the jihadist and white supremacist websites both used certain words to emphasize the victimization of the groups whose interests they claim to represent. The word "attack," which appeared most frequently on the jihadist website (6.38 times per page) was used to describe attacks against Afghanistan as "terrorism" and to raise awareness about attacks against the Quran and mosques. The word "destroy" (used 0.38 times per page) was employed in a similar manner, with reference to power stations and villages being destroyed. Even the word "Islam" (occurring 7.38 times per page) was often used in a manner that portrayed Muslims to be under attack; for instance:

- "The western and American print and electronic media are continuously spitting venom against Islam,"
- "The Muslims have already suffered too much of violence and tyranny but now the non-Muslim world plans to eliminate the Muslims once for all simply because strong Islam is something intolerable for the non-Muslim forces," and
- "The enemy has already declared a war against Islam".

In addition, the word "kill" (used 0.13 times per page) was further used to emphasize "the killing of innocent people" at the hands of the U.S.

Similarly, when the white supremacist website used words such as "attack," "bomb," "dead" and "destroy," it was to emphasize the victimization of "whites" at the hands of other "races". For example, "attack" (appearing 0.89 times per page) was used in the context of attacks on freedom of speech, various terrorist attacks against Americans, non-white immigrants as "attacks on freedom," and so on. The word "destroy" (occurring 0.57 times per page) was often used in the same way, describing how:

- "Mexicans will destroy America,"
- "Jewish heritage week will destroy American heritage," and
- "multiculturalist movement will destroy the fabric of White America."

The term "fight" (used 0.68 times per page) was also employed in a similar manner, with the white supremacist website noting that white individuals must "fight for the security and survival of [their] people," fight against "white racism," and fight against organized crime by fighting multiculturalism. Finally, the word "kill" was used to discuss various killings of "Whites" by "Blacks," although it was also used it in other contexts, including engaging in Holocaust denial by questioning the killing of Jewish people and discussing the killing of Saddam Hussein. Overall, a tendency emerged for these websites to use certain words in ways that emphasizes their role as victims with a sympathetic cause rather than as aggressors with a violent agenda. This allows websites to set the stage for encouraging action, further propaganda, and/or for recruitment purposes.

Some of the same words were also used from vastly different perspectives. For instance, each of the websites used the word "terrorist" to refer to American activities in the Middle East, but approached the issue from a different angle. Within the jihadist website, the word was used in reference to "U.S. terrorist attacks" against Islamic nations, to describe the "terrorist war against Muslim ummah ["community"]" launched by the U.S., to describe the "U.S. and its western allies' attack on Afghanistan as the worst kind of terrorism," and to describe America's allies as "terrorists." At the same time, the website seeks to dissociate the word terrorism from Muslims, by emphasizing that the "Noble Quran" does not preach terrorism and that mosques are not training grounds for terrorists, despite western "propaganda" to this effect.

The white-supremacist also uses the word "terrorism" largely to describe America's actions in the Middle East, criticizing the U.S. policy surrounding the "War against Terrorism." For instance, the website states that "Covert Operations are a huge part of the CIA [and] are simply state- sponsored terrorism," further arguing that "The only way we can end our War against Terrorism, is to end the US practice of conducing Terrorism under the guise of Covert Ops." The website also criticizes the use of detention centers such as Guantanamo Bay to hold so-called "terrorists" or suspected terrorists". The general consensus appears to be that the U.S. should focus more on the state of its nation within its borders rather than outside. As such, while both websites condemn American actions in the Middle East, they do so from different stand points and for different reasons.

Overall, the different types of websites can be differentiated by the different frequency of keywords used. However, both jihadist and white supremacist websites use various words for the same purpose (e.g., portraying themselves as victims) and use the same words for different purposes (e.g., attacking or defending the West or for recruitment or general discussion).

5 Discussion

The "National Strategy for Homeland Security" report in the U.S. emphasized that science and technology were important counter-terrorist tools [14]. It has been suggested that the use of information technology will increase national safety [13] by assisting in intelligence gathering through the collection and analysis of terrorism-related data [3]. This renders the creation of web-crawlers designed to detect and extract networks of extremist or terrorist web-sites a valuable enterprise, as these can build knowledge related to the content (i.e., group activities, recruitment processes, propaganda materials, etc.) of such websites and the affiliations of these groups.

This exploratory study used TENE, a specially designed web-crawler, to explore certain content aspects of two extremist websites (a white supremacist website and a jihadist website), with comparisons made to non-extremist websites (a counterterrorist website and a news website). It was found that all websites could be identified by certain keywords; for instance, the jihadist website used words such as "Allah," "Islam," and "Attack" at greater rates than the other websites. In addition, the extremist websites used language in specific ways, with words such as "attack," "destroy," and "dead" being used to emphasize the group's role as a victim. It should be stressed that the number of websites selected is small; as such, this project is entirely exploratory in nature, designed to provide preliminary information on how extremist websites might be identified by a web-crawler and how they might be used by terrorist or extremist groups.

Past studies have also explored the presence of extremist groups on the Internet and developed tools to collect extremist websites [3, 7, 17, 19, 20, 25]. Some organizations, including SITE institute, the Anti-Terrorism Coalition, and the Middle East Media Research Institute (MEMRI) have used manual analysis techniques to collect and monitor extremist websites. The Artificial Intelligence Lab uses automated processes for collection building. The Dark Web project has combined both manual and automated processes to build and analyze collections with the goal of combining the efficiency of automated techniques with the accuracy of manual ones. The TENE project seeks to extend past work on web-site-collection tools. It represents a return to automated processes for the purposes of efficiency; however, it also seeks to achieve a degree of accuracy similar to manual processes. This is done through the use of specific inclusion criteria within the web-crawler, such as keyword requirements. However, TENE is still in its early stages of development and is undergoing further modifications so that, among other things, it can properly differentiate between extremist and non-extremist websites.

To begin, the list of keywords will be refined to remove infrequently used words and to include other words that may capture a variety of extremist websites (e.g., Jihadic websites, neo-Nazi or white supremacist websites, eco-terrorist websites, southern separatist web-sites, etc.). The process of refining the keyword list will naturally require analysis that extends far beyond the initial examination of the two extremist websites for this project. In addition, including the keywords in a variety of languages (English, Arabic, French, etc.) is expected to improve the crawler's detection capacity. At present, TENE can identify and record the presence of images and videos; in the future, it may also be designed to recognize other website tools, such as chat rooms, forums, and donation options. Detection of interactive tools on websites is particularly important, as, for instance, [12] identifies a shift from websites of individual Jihadist groups to websites of pan-Jihadist forums.

A further refinement will involve the integration of semantics tool for the purpose of contextual analyses. Specific text mining and clustering techniques have been developed to uncover themes or topics (i.e., clusters of semantically related words) within sets of documents. Cucchiarelli et al. [5] used some of these tools to develop a content-based social network analysis (CB-SN) that they applied to various research and scientific networks. These tools include concept extraction and topic detection processes. Concept extraction identifies the relevant, domain-specific concepts after collecting textual information of interest (blogs, e-mails, articles, etc.). Topic detection uses a clustering algorithm to identify clusters of semantically similar concepts. In addition, researchers involved in the Dark Web project have begun exploring methods for sentiment and affect analysis in extremist websites [1]. Sentiment analysis distinguishes between text that contains positive and negative sentiments while affect analysis examines the emotions and moods expressed; in the Dark Web project, these analyses are achieved by looking at certain syntactic and stylistic features of websites [1]. By merging available tools with the web-crawler, emergent semantics and tone of language in extremist and terrorist websites can be more easily identified and information on popular topics can be extracted.

One of the ancillary benefits of TENE is its sustainability. This tool has already demonstrated its benefits in relation to the investigation of child exploitation on the Internet. Following its adaptation to the study of extremism, it will lend itself to further substantive applications, such as the funding of terrorism and the spread of propaganda. We also envision TENE as an integral part of a broader strategy to disrupt extremist networks.

References

1. Chen, W.: Dark Web: Exploring and Data Mining the Dark Side of the Web. Springer, New York (2012)
2. Chen, H., Chung, W., Qin, J., Reid, E., Sageman, M., Weinmann, G.: Uncovering the dark web: a case study of Jihad on the Web. J. Am. Soc. Inf. Sci.Technol. **59**(8), 13471359 (2008)
3. Chen, H., Qin, J., Reid, E., Chung, W., Zhou, Y., Xi, W., Lai, G., Elhourani, T., Bonillas, A., Wang, F.Y., Sageman, M.: Collecting and analyzing the presence of domestic and international terrorist groups on the Web. In: Proceedings of the 7th Annual IEEE Conference on Intelligent Transportation Systems. The dark web portal (2004)
4. Conway, M.: Reality bites: Cyberterrorism and terrorist use' of the Internet. http://www.firstmonday.org/Issues/issue7_11/conway/index.html (2002)
5. Cucchiarelli, A., D'Antonio, F., Velardi, P.: Semantically interconnected social networks. Soc. Netw. Anal. Min. **2**(1), 69–95 (2012)
6. Frank, R., Westlake, B., Bouchard, M.: The structure and content of online child exploitation networks. In: Proceedings of the Tenth ACM SIGKDD Workshop on Intelligence and Security Informatics '04 (2010)
7. Institute for Security Technology Studies, Technical Analysis Group: Examining the cyber capabilities of Islamic Terrorist Groups. http://www.ists.dartmouth.edu/ (2004)
8. Jacobson, M.: Terrorist financing and the Internet. Stud. Confl. Terror. **33**(4), 353–363 (2010)
9. Jiang, C., Coenen, F., Sanderson, R., Zito, M.: Text classification using graph mining based feature extraction. In: Proceedings of the SGAI '09: International Conference on Artificial Intelligence, London (2009)
10. Joffres, K., Bouchard, M., Frank, R., Westlake, B.: Strategies to disrupt online child pornography networks. In: Proceedings of the 11th ACM SIGKDD Workshop on Intelligence and Security Informatics. 163–170 (2011)
11. Kamruzzaman, S.M., Farhana, H., Ahmen, R.H.: Text classification using data mining. In: Proceedings of International Conference on Information and Communication Technology in Management '05, Multimedia University, Malaysia (2009/10)
12. Musawi, M.A.: Cheering for Osama: how Jihadists use internet discussion forums. Quillan Foundation. http://www.quilliamfoundation.org/images/stories/pdfs/cheering-for-osama.pdf(2010)
13. National Research Council: Making the Nation Safer: The Role of Science and Technology in Countering Terrorism. National Academy of Sciences, Washington DC (2002)
14. Office of Homeland Security: National strategy for homeland security. Office of Homeland Security, Washington DC (2002)
15. Sebastiani, F.: Machine learning in automated text categorization. ACM Comput. Surv. **34**(1), 1–47 (2002)
16. Technical Analysis Group: Examining the cyber capabilities of Islamic Terrorist Groups. Institute for Security Technology Studies at Dartmouth College, Hanover (2004)
17. Thomas, T.L.: Al Qaeda and the Internet: the danger of cyberplanning'. Parameters **33**, 112–123 (2003)
18. Tong, S., Koller, D.: Support vector machine active learning with applications to text classification. J. Mach. Learn. Res. **2**, 45–66 (2002)
19. Tsfati, W., Weimann, G.: http://www.terrorism.com: Terror on the Internet. Stud. Confl. Terror. **25**(3), 317–332 (2002)
20. Weimann, G.: http://www.terror.net: How modern terrorism uses the Internet. Special report, US Institute of Peace. http://www.usip.org/pubs/specialreports/sr116.pdf (2004)
21. Weimann, G.: Terror on the Internet: The New Arena, The New Challenges. United States Institute of Peace, Washington DC (2006)
22. Westlake, B., Bouchard, M., Frank, R.: Finding the key players in online child exploitation networks. Policy Internet **3**(2), 104 (2011)
23. Xu, J., Chen, H., Zhou, Y., Qin, J.: On the topology of the dark web of terrorist groups. Lect. Notes Comput. Sci. **3975**, 367–376 (2006)

24. Zhou, Y., Qin, J., Lai, G., Reid, E., Chen, H.: Building knowledge management system for researching terrorist groups on the Web. In: Proceedings of the 11th Americas Conference on Information Systems '05, Omaha, NE, USA (2005)
25. Zhou, Y., Reid, E., Qin, J., Chen, H., Lai, G.: U.S. Extremist Groups on the Web: link and content analysis. IEEE Intell. Syst. **20**(5), 44–51 (2005)

Chapter 12
Sampling Emerging Social Behavior in Facebook Using Random Walk Models

C. A. Piña-García and Dongbing Gu

Abstract It has long been recognized that random walk models apply to a great diversity of situations such as: economics, mathematics and biophysics; current trends about Open Social Networks require new approaches for analyzing material publicly accessible. Thus, in this chapter we examine the potential of random walks to further our understanding about monitoring Social Behavior, taking Facebook as a case study. Although most of the work related to random walk models is traditionally used to generate animal movement paths, it is also possible to adapt classic diffusion models into exploratory algorithms with the aim to improve the ability to search under a complex environment. This algorithmic abstraction provides an analogy for a dissipative process within which trajectories are drawn through the virtual nodes of Facebook.

1 Introduction

The problem of monitoring social behavior including their security implications have become in a paramount task for security agencies around the world. Therefore, sampling and monitoring social networks such as Facebook is a major application area that can be seen as means to detect emerging social behavior in a virtual environment. This kind of uncertainty about what is actually happening around us has led us to develop control mechanisms able to cope with information flows generated by the Open Social Networks (OSNs).

C. A. Piña-García (✉) · D. Gu
School of Computer Science and Electronic Engineering, University of Essex, Wivenhoe
Park, Colchester CO4 3SQ, UK
e-mail: capina@essex.ac.uk

D. Gu
e-mail: dgu@essex.ac.uk

V. K. Mago and V. Dabbaghian (eds.), *Computational Models of Complex Systems*,
Intelligent Systems Reference Library 53, DOI: 10.1007/978-3-319-01285-8_12,
© Springer International Publishing Switzerland 2014

There is a growing tendency to exchange information through the OSNs e.g., messages, emails, pictures, videos and breaking news [1]. This property of the OSNs has led us to interact with social media systems, losing in some degree our privacy rights. Under such a situation, many users deliberately share all their information with the rest of the world by just uploading or posting something on any OSN. In any attempt to monitor emerging social behavior in Facebook it is necessary to distinguish what kind of information sources we are interested i.e., messages, pictures, locations, events, users, etc. All this information can be collected for further examination.

In this study we consider the Facebook structure as a whole which is characterized by a graph, where all its entities such as vertices and edges are modeled from a Graph Theory point of view [2]. However, it should be stressed that features describing the network structure i.e., node degree, number of common friends, shortest path length will not be considered in this study.

In Sect. 2 we present a review of the literature on four different random movement models together with their statistical properties. In the second part of the chapter we present an exploratory system that searches for "key events" and how to use optimal search strategies in Facebook. We also review the paths generated by the random walk models. The third part of the chapter addresses the designed set of experiments to evaluate our assumptions and the exploration feature for sampling emergent social behavior on Facebook. Finally, we report representative results on real time data. We further demonstrate that random walk models are particularly suitable for searching large-scale social networks.

2 Random Walk Models

Traditionally, random walk models have addressed the question of what is the best statistical strategy to adapt in order to search efficiency for randomly located objects [3]. In this regard, we have consider a set of random walks that depends on the probability distribution of step lengths taken by a random walker. Thus, the question of how direction of movement affects the path of the walker provides us with various random movement models to be tested. Generally, it is possible to identify a set of diffusion models where movement paths present idealized patterns which captures some of the essential dynamics of animal foraging.

According to [4] a general property arises from the using of optimal strategies: the more heterogeneous the target distribution, the more probable is that after an encounter a search starts with a nearby target. Furthermore, it is important to note that there is not a universal solution to searching problems i.e., this kind of problems are sensitive to initial conditions. Thus, bio-inspired movement paths can be optimal strategies for searching randomly targets, this kind of models show interesting properties e.g., scale invariance (fractal property) and super diffusion at all scales [5].

A foraging task could be defined as an extensive searching until a target is found, followed by an intensive grazing period of limited length that is generally an efficient strategy in all types of environments [7]. Generally, random walk models can be considered as processes with an associated high degree of uncertainty. However, an important feature of these models is that they can provide a behavioral flexibility to adapt in different scenarios [8].

A random walk model is a formalization of the intuitive idea of taking successive steps, each in a random direction. Thus, they are simple stochastic processes consisting of a discrete sequence of displacement events (i.e., move lengths) separated by successive reorientation events (i.e., turning angles) [8]. The influence of reorientations is another key concept that we must take into account, in particular we need to consider how this degree of directional memory (i.e., persistence) in the walk has led the explorer to optimize encounter rates [9]. There are two different features related to turning angle distributions:

1. The shape (relative *kurtosis*[1]) of the angular distribution and
2. The correlations between successive relative orientations (directional memory).

This variability in the shape of the turning angle distributions can appear only at limited spatio-temporal scales [9] i.e., after a certain amount of time some random walk models tend to settle down as a *Gaussian* distribution [7]. It should be noted that the simplicity of random walk models is methodologically attractive for using in several scenarios [10]. Thus, improving the ability to search involves the selection of a specific set of "rules of search" that enhances the probability of finding unknown located items [11]. It seems that different optimal solutions arise by merely embracing different random strategies [6]. Therefore, in any given environment there might be a range of search strategies that can be successful, and individuals may differ in the search strategy used [11].

In the present work we focus our attention on those particular cases where the efficiency of searching and sampling is fully determined by the statistical features of turning angle distributions. In this chapter we consider the problem of a social explorer searching for specific information inside a OSN over a short-time period. Thus, it raises the question of whether a social explorer is able to cope with an OSN (Facebook).

The statistical properties of these random walks are as follows:

- **Lévy Walk:** Lévy walks are a class of stochastic processes based on the Lévy-stable distribution. The stochastic processes arising from such distributions are tightly related to anomalous diffusion phenomena. Furthermore, in Lévy walks, a explorer must move along the trajectory and the time to complete a jump is involved [8]. From the statistical point of view, Lévy walks are characterized by a distribution function $P(l_j) \sim l_j^{-\mu}$ with $1 < \mu \leq 3$ where l_j is the flight length and the symbol $\sim$ refers to the asymptotic limiting behavior as the relevant quantity goes to infinity or zero [3]. In a Lévy walk the turning angles are usually not directly considered (turning angles are uniform on the unit circle). For the special case when $\mu \geq 3$ a Gaussian distribution arises due to the central limit theorem [12]. On the other hand, when $\mu \leq 1$ the probability distribution cannot be normalized.

[1] Kurtosis is the degree of peakedness of a distribution. Higher kurtosis means more of the variance is the result of infrequent extreme deviations.

The exponent of the power-law is named the Lévy index (μ) and controls the range of correlations in the movement, introducing a family of distributions, ranging from *Brownian motion* ($\mu > 3$) to straight-line paths ($\mu \rightarrow 1$) [11]. Figure 1a shows a plot of a simulated Lévy walk.

- **Brownian Motion:** The classic diffusion model known as *Brownian motion* it is a simple strategy uncorrelated and unbiased. Uncorrelated means that the direction of movement is completely independent of the previous direction and unbiased means that there is no preferred direction: the direction moved at each step is completely random. This model presents some features such as: explorations over short distances can be made in much shorter times than explorations over long distances. The random walker tends to explore a given region of space rather thoroughly. It tends to return to the same point many times before finally wandering

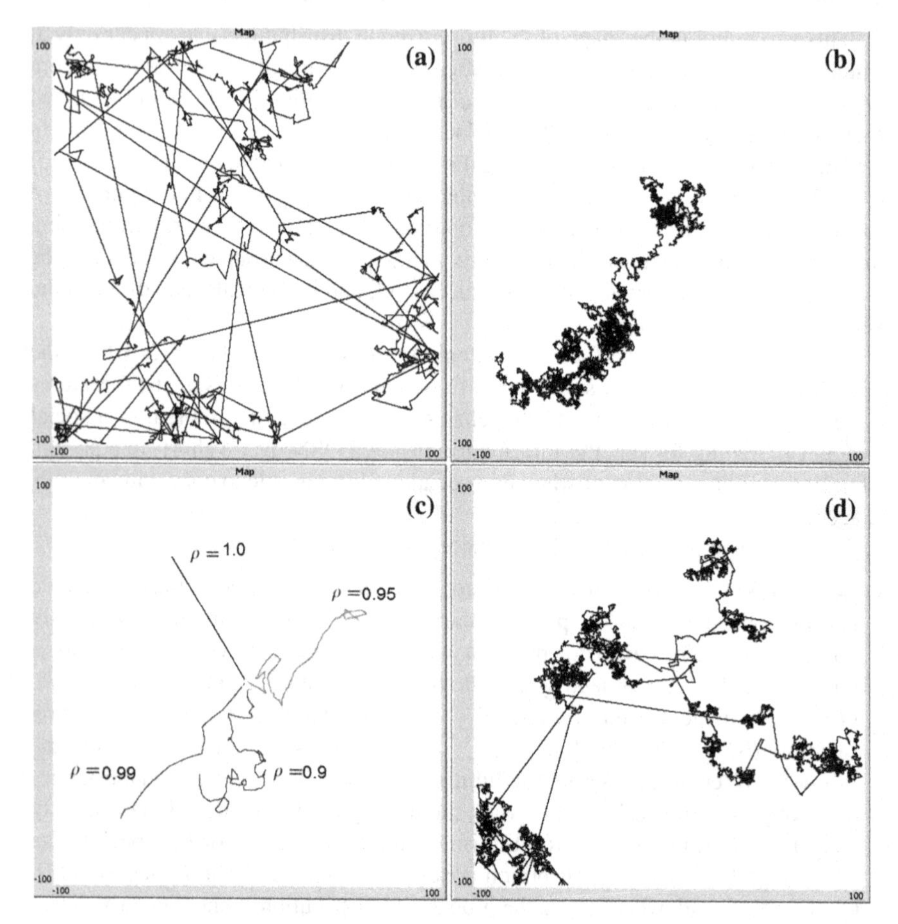

Fig. 1 Plots showing simulated random walks. **a** Lévy walk with $\mu = 2$; **b** Brownian walk with $\mu = 3$; **c** Correlated Random Walk (CRW) with $0 \leq \rho \leq 1$ and **d** adaptive or composite strategy switching between Lévy and brownian motion

away. It chooses new regions to explore blindly. In addition, the random walker has no any tendency to move toward regions that it has not occupied before. It has absolutely no inkling of the past and finally its track does not fill up the space uniformly [13]. Figure 1b shows a plot of a simulated Brownian walk.

- **Correlated Random Walk (CRW):** Correlated random walks (CRWs) involve a correlation between successive step orientations, where movement paths show *persistence*. This produces a local directional bias: each step tends to point in the same direction as the previous one [7]. The CRW consists of a series of discrete steps of length l_j and direction θ_j. The length l_j of the jth move and the turning angle $\phi_j = \theta_{j+1} - \theta_j$ are assumed to be random variables with no autocorrelation or cross-correlation (and no correlation between step length and step direction) [7]. Furthermore, the simplest way to incorporate directional persistence into a random walk model is introducing correlations (i.e., memory effects) between successive random walk steps. Thus, the trajectories generated by correlated random walk models appear more similar to the empirical data than those generated by uncorrelated random walks (e.g., Brownian motion). This model controls directional persistence via the probability distribution of turning angles, combining persistence with a preferred direction. This strategy is characterized by using a wrapped Cauchy distribution (WCD) [14] for the turning angles

$$\theta = \left[2 \times \arctan\left(\frac{(1-\rho) \times \tan(\pi \times (r - 0.5))}{1+\rho}\right)\right], \tag{1}$$

where ρ is the shape parameter ($0 \leq \rho \leq 1$) and r is an uniformly distributed random variable $r \in [0, 1]$. Directional persistence is controlled by changing the shape parameter of the WCD (ρ). Thus, for $\rho = 0$ we obtain an uniform distribution with no correlation between successive steps (Brownian motion), and for $\rho = 1$, we get a delta distribution at 0°, leading to straight-line searches. Figure 1c shows a plot of a simulated CRW.

- **Adaptive or Composite strategy:** In some cases when purely random searching models become less effective, the explorer must attempt to move in such a way so as to optimize their chances of locating targets by increasing the chances of covering a given area [11]. A random search model where the explorer can change its behavior depending on the environmental conditions is known as an *adaptive* search [4]. Thus, a composite strategy or an adaptive random walk consists of an explorer undergoing an extensive search[2] (in this study: a Lévy walk) until a target is encountered, at which point the explorer changes to Brownian motion to undergo an intensive search [15–17]. This kind of model switches between a Brownian motion and a Lévy walk according to a biological oscillator used by Nurzaman et al. [18]. This switch is based on environmental changes (encounter rate) sensed by the forager. We use an adaptive switching behavior defined in [18, 19]. Specifically, we compute $P(t) = \exp(-z(t))$ with a conditional function

[2] An extensive search consist in long steps that improve the searching efficiency by using super-diffusive phases, e.g., ballistic motion and Lévy walks.

where if $P(t) = 1$, then a Brownian motion is triggered. Otherwise, a Lévy walk is used as a default behavior. Figure 1d shows a plot of a simulated composite strategy.

3 Emerging Social Behavior in Facebook

We develop a social explorer that searches for "key events" and provides a quick mechanism to monitor Facebook, tracking the ebb and flow of social media as an innovative approach. It should be stressed that most of the emerging social events are regularly reported by any social network e.g., Facebook, Twitter and Foursquare. This consideration implies that *Social Media* has become a primary source of information about events and developing situations.

Velasco and Keller in [20, 21] respectively, pointed out that the main goal is to view publicly available open source and non-private social data that is readily available on the Internet. Our application is not focused on specific persons or protected groups. It is mainly focused on words related to events and activities related on social behavior or potential threats. Examples of these words include lock down, bomb, suspicious package, white powder, active shoot, etc. This technique intends to capture Facebook users related to a specific keyword or key event with the aim to display an activity plot (see Fig. 2) showing popular or emerging events according to their attending list obtained from Facebook. These plots present a graphical random distribution based on any of the previously mentioned random walk models presented in Sect, 2. Thus, our study can be considered as a way of quickly getting an idea about whether a status update has been posted by someone that potentially could be involved in a particular event.

One of the main problems highlighted in social media is about location and mapping "bad actors" and analyzing their movements, vulnerabilities, limitations, and possible adverse actions [20]. Thus, the use of social media to target specific users or groups of users becomes in a high priority plan for security agencies around the world. It seems reasonable to assume that security agencies could be alerted if some specific search produces evidence of breaking events, incidents, and emerging social behavior. These breaking events are graphically symbolized by clusters of users located around their respective activities on Facebook (see Fig. 2).

4 How to Use Optimal Search Strategies in Facebook

Initially, we consider the Facebook social network as an unknown complex environment modeled by an undirected graph $G = (V, E)$, where V is a set of vertices assuming the role of nodes (users) and E assuming two different roles. On one hand, the first role is determined by those paths traced by the social explorer. On the other hand, the second role is determined by all the edges linking a key event to its attending

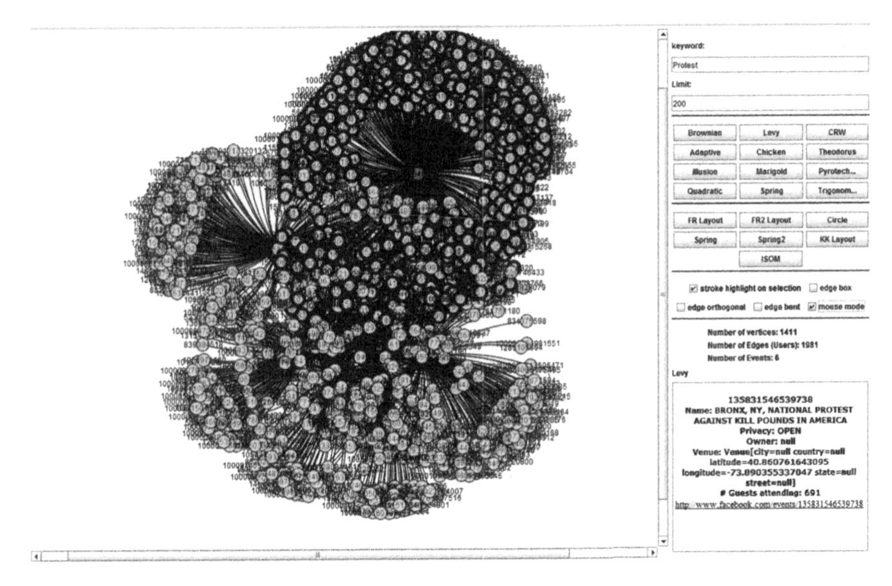

Fig. 2 An example plot showing a random distribution of nodes/users on Facebook. This distribution is based on a keyword as a parameter of search. Users are displayed in various clusters according to a particular attractor, in this case a key event

list of users. In this work we consider two stages: (1) analysis of movement pattern and (2) linkage between a key event and its corresponding attending list.

4.1 Analysis of Movement Pattern

This stage consists of a social explorer that carries out an extensive search based on a keyword until a target node is found. The searching mechanism is based on one of the random walk models mentioned in Sect. 2. Thus, we propose a single social explorer to interact with Facebook with the intention to obtain publicly available information from this network.

Facebook is a large-scale social network that is continually changing over time [22]. Therefore, Facebook is essentially unknowable. But at the same time, it does not need to be known in detail [1, 23].

A sampling example movement path based in random walk models explained in Sect. 2 is given in Fig. 3. In this Figure we show our proposed process of a random walk carried out in Facebook. Firstly, a sample is obtained from a RW model. In this case the information is provided as a set of key event nodes. It should be noted that these key event nodes have a list of attending users (not displayed at this stage). Thus, all nodes are placed randomly on the graphical user interface (GUI), see Fig. 3a.

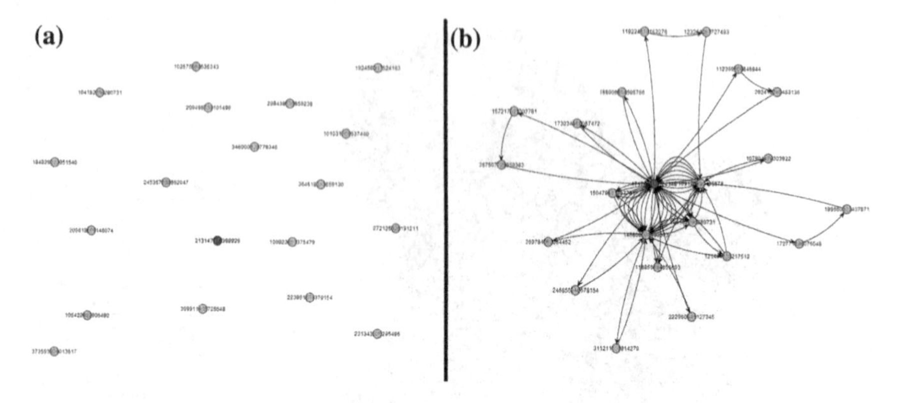

Fig. 3 Representative examples of a movement path based on a random walk model displaying only key event nodes (**a**), and a movement path linking all the key event nodes to their respective edges forming a statistically generated pattern (**b**)

In Fig. 3b, the key event nodes are linked to each other according to a specific random walk model. Therefore, a movement pattern is statistically generated by a stochastic process and depicted on the GUI. It is important to mention that this kind of links are merely conceptual i.e., we are taking into consideration the underlying processes that generate them. As described earlier the initial stage works as a tracking and monitoring system based on initial conditions determined by a keyword.

4.2 Linkage Between a Key Event and Its Corresponding List of Attending

In this second stage, each user who is attending to a particular event is directly linked to its event, i.e., a key event becomes in a kernel for all the surrounding users (see Fig. 4a). This linkage process continues for every single event that belongs to the random sample. Thus, each event is linked to a group of users forming various clusters of nodes presented in the GUI. At this point, it is possible to distinguish between two types of elements: events and users; and at the same time it is also feasible to identify which user is attending to what event, see Fig. 4b.

5 Methods

We have designed a set of experiments to evaluate our assumptions about improving the ability of searching and sampling in Facebook using random walk models. We investigated the influence of random walks on the probability of successful social network exploration. We also investigated the exploration feature for sampling emergent

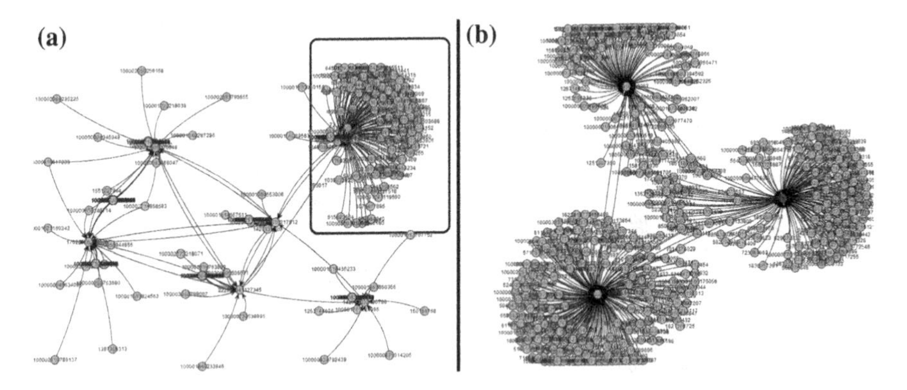

Fig. 4 Representative examples of (**a**) the early stage of the structural pattern forming a kernel surrounded by collections of nodes, and (**b**) the final stage showing a set of clusters linking all the key event nodes to their respective attending group of users

Table 1 List of twelve proposed keywords used as search parameters on the social network environment (Facebook)

1. Gangs	4. Crisis	7. Anonymous	10. Government
2. Recall	5. Bomb	8. Afghanistan	11. Revolution
3. Hack	6. Riot	9. Syria	12. Justice

social behavior through viewing publicly available social data. We used our social explorer for searching a list of twelve proposed keywords[3] (see Table 1). The investigations were conducted using the following metrics: number of vertices, number of edges, number of events and elapsed time.

6 Results

We report representative results from experiments conducted in a social network environment using four strategies. Across all random walk models, the number of nodes/vertices explored (V), edges explored (E) and events examined in the *Brownian motion* case were greater than the rest of strategies (see Fig. 5a, b). Our results agree with those reported in [6], where it is argued that a Brownian motion follows a *Gaussian* distribution. Thus, a redundant search is able to explore or examine more nodes (users or events) even in a OSN scenario.

Results regarding the elapsed time reflect assumptions about complexity of an algorithm from the statistical point of view, e.g., a CRW spent more time than the rest of strategies probably due to how the turning angles are computed (see Eq. 1). In contrast, a Brownian motion and a Lévy walk spent less time searching in Facebook

[3] This list of keywords was mainly suggested by public preferences and current breaking news.

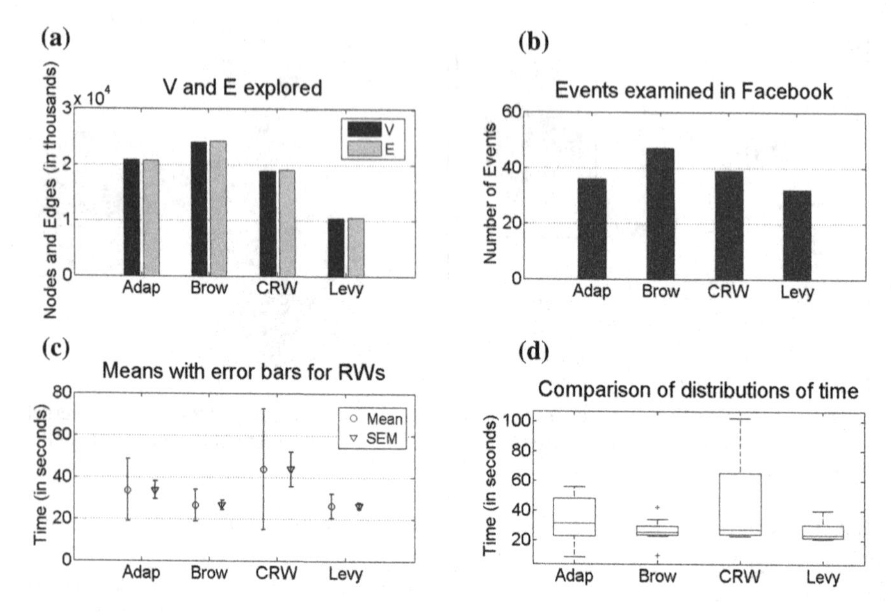

Fig. 5 **a** Total of vertices explored **V** and edges explored **E.** **b** Total of events examined in Facebook. **c** Means and SEMs bars corresponding to the elapsed time and **d** variability in the elapsed time using four different strategies

(see Fig. 5c). Brownian motion and a Lévy walk presented reduced variability in their elapsed times. This reflects that these strategies showed the least variable time response (see Fig. 5d). Finally, we found that our reported measures showed statistical differences when mean and standard error mean (SEM) were compared in Fig. 5c. Our results suggest that the average time of sampling is fully acceptable for using these algorithms on Facebook.

Exploratory experiments were also completed with the aim to compare the ability to search using the list of twelve keywords provided in Table 1. It should be stressed that results regarding the elapsed time could be associated with the response capability of Facebook servers (see Fig. 6).

7 Conclusions

We have completed a comprehensive study of Social Networks exploration using random walk models. Our results agree with classic studies in the literature where the influence of turning angles for optimizing encounter rates is used on non-oriented searches [3, 9]. Our comparison of the performance of several search strategies on Facebook suggests that searching and sampling the content of Facebook can be used as a simple methodology for network exploration. It should be stressed that we are

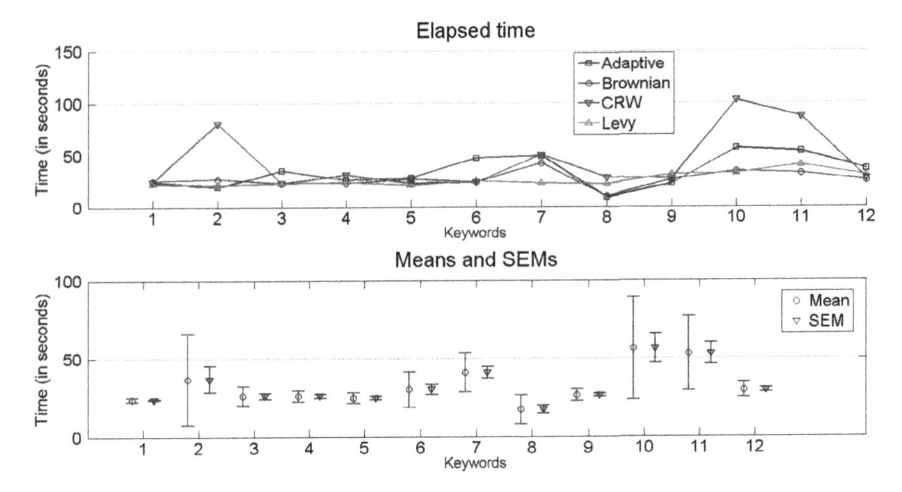

Fig. 6 Elapsed time of the 12 keywords during an exhaustive exploration through Facebook. Means and SEMs bars corresponding to the elapsed time are also indicated below

using a real time social explorer and our results should not be considered only in a quantitative sense but also on a qualitative sense related to a conceptual modeling. The main contribution of this work is that it compares the performances of four different techniques with the aim to monitor social behaviors in Facebook in an informative way.

Future work in this area should consider different random walk models provided by natural mechanisms or artificial techniques, these kind of models should examine all trajectories generated with the aim to take decisions about which search strategy to follow. We can also suggest some areas of potential application for our study, e.g., detecting and monitoring threats; taking into account that privacy laws have been passed and that actually enable monitoring social networks even when settings are set to private [24].

Acknowledgments The authors thank the reviewers of this chapter for their useful comments. Mr. Piña-García has been supported by the Mexican National Council of Science and Technology (CONACYT), through the program "Becas para estudios de posgrado en el extranjero" (no. 213550).

References

1. Traud, A.L., Mucha, P.J., Porter, M.A.: Social structure of facebook networks. Physica A, (2011)
2. Gjoka, M., Kurant, M., Butts, C.T., Markopoulou, A.: Walking in facebook: A case study of unbiased sampling of osns. In: INFOCOM, 2010 Proceedings IEEE, IEEE, pp. 1–9 (2010)
3. Viswanathan, G.M., Buldyrev, S.V., Havlin, S., Da Luz, M.G.E., Raposo, E.P., Stanley, H.E.: Optimizing the success of random searches. Nature **401**(6756), 911–914 (1999)

4. Reynolds, A.M., Bartumeus, F.: Optimising the success of random destructive searches: Lévy walks can outperform ballistic motions. J. Theor. Biol. **260**(1), 98–103 (2009)
5. Plank, M.J., Codling, E.A.: Sampling rate and misidentification of Lévy and non-Lévy movement paths. Ecology **90**(12), 3546–3553 (2009)
6. Viswanathan, G.M., da Luz, M.G.E., Raposo, E.P., Stanley, H.E.: The Physics of Foraging: An Introduction to Random Searches and Biological Encounters. Cambridge University Press, Cambridge (2011)
7. Codling, E.A., Plank, M.J., Benhamou, S.: Random walk models in biology. J. R. Soc. Interface **5**(25), 813 (2008)
8. Bartumeus, F.: Lévy processes in animal movement: an evolutionary hypothesis. Fractals London **15**(2), 151 (2007)
9. Bartumeus, F., Catalan, J., Viswanathan, G.M., Raposo, E.P., da Luz, M.G.E.: The influence of turning angles on the success of non-oriented animal searches. J. Theor. Biol. **252**(1), 43–55 (2008)
10. Zollner, P.A., Lima, S.L.: Search strategies for landscape-level interpatch movements. Ecology **80**(3), 1019–1030 (1999)
11. Bartumeus, F., Da Luz, M.G.E., Viswanathan, G.M., Catalan, J.: Animal search strategies: a quantitative random-walk analysis. Ecology **86**(11), 3078–3087 (2005)
12. Anderson, D.R., Sweeney, D.J., Williams, T.A.: Introduction to Statistics: Concepts and Applications. West Publishing, St. Paul (1994)
13. Berg, H.C.: Random Walks in Biology. Princeton University Press, Princeton (1993)
14. Kent, J.T., Tyler, D.E.: Maximum likelihood estimation for the wrapped cauchy distribution. J. Appl. Stat. **15**(2), 247–254 (1988)
15. Bénichou, O., Coppey, M., Moreau, M., Moreau, M., Suet, P.H.: Optimal search strategies for hidden targets. Phys. Rev. Lett. **94**(19), 198101 (2005)
16. Bénichou, O., Coppey, M., Moreau, M., Suet, P.H., Voituriez, R.: A stochastic theory for the intermittent behaviour of foraging animals. Physica A **356**(1), 151–156 (2005)
17. Bénichou, O., Coppey, M., Moreau, M., Voituriez, R.: Intermittent search strategies: when losing time becomes efficient. EPL (Europhys. Lett.), **75** 349 (2006)
18. Nurzaman, S.G., Matsumoto, Y., Nakamura, Y., Shirai, K., Koizumi, S., Ishiguro, H.: An adaptive switching behavior between levy and Brownian random search in a mobile robot based on biological fluctuation. In: 2010 IEEE/RSJ International Conference on Intelligent Robots and Systems (IROS), pp. 1927–1934, IEEE (1927)
19. Pina-Garcia, C., Gu, D., Hu, H.: A composite random walk for facing environmental uncertainty and reduced perceptual capabilities. In: Intelligent Robotics and Applications, pp. 620–629 (2011)
20. Velasco, R.: webpage on Federal Business Opportunities (FBO). https://www.fbo.gov/spg/DOJ/FBI/PPMS1/SocialMediaApplication/listing.html
21. Keller, J.: How The CIA Uses Social Media to Track How People Feel. [Online] Available: http://www.theatlantic.com/technology/archive/2011/11/how-the-cia-uses-social-media-to-track-how-people-feel/247923/#.Trod98BsW3E.facebook
22. Tang, J., Musolesi, M., Mascolo, C., Latora, V.: Temporal distance metrics for social network analysis. In: Proceedings of the 2nd ACM Workshop on Online Social Networks, pp. 31–36. ACM (2009)
23. Ugander, J., Karrer, B., Backstrom, L., Marlow, C.: The anatomy of the facebook social graph. Arxiv, preprint arXiv:1111.4503 (2011)
24. Gellman, R.: Perspectives on privacy and terrorism: all is not lost yet. Gov. Inf. Q. **19**(3), 255–264 (2002)

Printed in the United States
By Bookmasters